Mario Richter

Setd6: A novel histone lysine mono-methyltransferase

Mario Richter

Setd6: A novel histone lysine mono-methyltransferase

Setd6 is a novel histone lysine monomethyltransferase which will be important for the understanding of signaling to chromatin

Südwestdeutscher Verlag für Hochschulschriften

Impressum/Imprint (nur für Deutschland/ only for Germany)
Bibliografische Information der Deutschen Nationalbibliothek: Die Deutsche Nationalbibliothek
verzeichnet diese Publikation in der Deutschen Nationalbibliografie; detaillierte bibliografische
Daten sind im Internet über http://dnb.d-nb.de abrufbar.
Alle in diesem Buch genannten Marken und Produktnamen unterliegen warenzeichen-, marken-
oder patentrechtlichem Schutz bzw. sind Warenzeichen oder eingetragene Warenzeichen der
jeweiligen Inhaber. Die Wiedergabe von Marken, Produktnamen, Gebrauchsnamen,
Handelsnamen, Warenbezeichnungen u.s.w. in diesem Werk berechtigt auch ohne besondere
Kennzeichnung nicht zu der Annahme, dass solche Namen im Sinne der Warenzeichen- und
Markenschutzgesetzgebung als frei zu betrachten wären und daher von jedermann benutzt
werden dürften.

Verlag: Südwestdeutscher Verlag für Hochschulschriften Aktiengesellschaft & Co. KG
Dudweiler Landstr. 99, 66123 Saarbrücken, Deutschland
Telefon +49 681 37 20 271-1, Telefax +49 681 37 20 271-0, Email: info@svh-verlag.de
Zugl.: Wien, Universität Wien, Diss., 2009

Herstellung in Deutschland:
Schaltungsdienst Lange o.H.G., Berlin
Books on Demand GmbH, Norderstedt
Reha GmbH, Saarbrücken
Amazon Distribution GmbH, Leipzig
ISBN: 978-3-8381-0575-8

Imprint (only for USA, GB)
Bibliographic information published by the Deutsche Nationalbibliothek: The Deutsche
Nationalbibliothek lists this publication in the Deutsche Nationalbibliografie; detailed
bibliographic data are available in the Internet at http://dnb.d-nb.de.
Any brand names and product names mentioned in this book are subject to trademark, brand or
patent protection and are trademarks or registered trademarks of their respective holders. The
use of brand names, product names, common names, trade names, product descriptions etc.
even without a particular marking in this works is in no way to be construed to mean that such
names may be regarded as unrestricted in respect of trademark and brand protection legislation
and could thus be used by anyone.

Publisher:
Südwestdeutscher Verlag für Hochschulschriften Aktiengesellschaft & Co. KG
Dudweiler Landstr. 99, 66123 Saarbrücken, Germany
Phone +49 681 37 20 271-1, Fax +49 681 37 20 271-0, Email: info@svh-verlag.de

Copyright © 2009 by the author and Südwestdeutscher Verlag für Hochschulschriften
Aktiengesellschaft & Co. KG and licensors
All rights reserved. Saarbrücken 2009

Printed in the U.S.A.
Printed in the U.K. by (see last page)
ISBN: 978-3-8381-0575-8

DISSERTATION

Setd6:

Characterization of a novel histone lysine mono-methyltransferase

angestrebter akademischer Grad

Doktor der Naturwissenschaften (Dr. rer.nat.)

Verfasserin / Verfasser:	Mario Steffen Richter
Matrikel-Nummer:	0648358
Dissertationsgebiet (lt. Studienblatt):	Molekulare Biologie
Betreuerin / Betreuer:	Prof. Dr. Thomas Jenuwein
Wien, am 05 Februar 2009	

TABLE OF CONTENTS
1. Summary 4
    1.1. Zusammenfassung 6
2. Introduction 9
    2.1. Epigenetics 9
    2.2. Epigenetic regulation by histone lysine methylation 11
    2.3. Interplay between histone modifications 15
    2.4. Non-histone lysine methylation 18
3. Setd6 is a histone lysine mono-methyltransferase interacting with remodeling activities and components of the Wnt signaling pathway. 21
    3.1. Abstract 21
    3.2. Introduction 22
        3.2.1. SET domain containing enzymes 22
    3.3. Results 25
        3.3.1. Setd3, Setd4 and Setd6 have a common domain architecture 25
        3.3.2. Setd6 localizes to the nucleus in murine NIH 3T3 cells 27
        3.3.3. Setd6 is transcribed in adult tissues and throughout mouse development ...28
        3.3.4. Setd6 methylates histone H1, H2A, H3 and H4 in vitro 30
        3.3.5. Setd6 is a histone lysine mono-methyltransferase 31
        3.3.6. Setd6 methylates H1K26, H2AK5, H3K14, H4K5 and H4K12 33
        3.3.7. Reduced Setd6 protein level in cells leads to decreased methylation of Setd6 target residues 36
        3.3.8. Staining patterns for H1K26, H2AK5, H3K14, H4K5 and H4K12 39
        3.3.9. Setd6 is involved in transcriptional control 42
        3.3.10. Setd6 interacts with H1 and components of the Wnt pathway 45
    3.4. Discussion 48
    3.5. Additional results for Setd6 54
        3.5.1. An increased level of Setd6 leads to the formation of micro nuclei 54
        3.5.2. Setd6 methylates non-histone proteins 56
    3.6. Discussion of additional results 60
4. Lysine methylation on p53 61
    4.1. Introduction 61
        4.1.1. Post-translational modifications of p53 61
    4.2. Repression of p53 activity by Smyd2-mediated methylation 64
    4.3. p53 is regulated by the lysine demethylase LSD1 78
    4.4. Discussion 90
5. Materials and Methods 92
6. Appendix 102
    6.1. Buffer composition: 102
    6.2. Media composition and concentration of antibiotics: 106
    6.3. Peptide sequences: 107
    6.4. Primer sequences: 108
    6.5. Plasmid maps: 109
    6.6. Dot-blots 116
7. References 120
8. Acknowledgments 126

1. Summary

In the field of epigenetics many mechanism have been uncovered that help to understand how different types of cells in an organism having the same genetic information can have distinct expression profiles. Those mechanisms include the posttranslational modification of the histones in the chromatin fiber, methylation of the DNA, non-coding RNAs acting in trans, incorporation of histone variants and defined micro environments in so called nuclear territories (Allis 2007). A field less explored is how extracellular signals can travel from the cell surface to the nucleus and then alter the transcriptional profile of a cell inducing e.g. differentiation, migration or apoptosis. In the first part of this thesis we characterize a novel nuclear histone lysine mono-methyltransferase (Setd6) that modifies residues in histone H2A, H3 and H4 that have so far only been characterized in their acetylated form. From the approximately 50 histone methyltransferase containing a SET domain many different activities have been characterized but mono-methylating enzymes are underrepresented, while di- and tri-methylating enzymes have been identified several times even targeting the same histone residues. This might reflect the fact that the mono-methylated state seems to be the initial step maybe occurring even before the incorporation of free histones into the nucleosome. Even more interestingly the residues modified by Setd6 are the same residues acetylated by the Tip60 (60 kDa Tat interactive protein) complex. By immunoprecipitation we found interactions of Setd6 with remodeling enzymes from this complex. We propose an antagonistic role for the Setd6 and Tip60 complex. Tip60 has been linked to pluripotency in embryonic stem cells (Fazzio et al. 2008) and the role of Setd6 in this process will be studied in future experiments. We also identified β-Catenin as an interacting protein, being the intracellular key regulator of the Wnt signaling pathway and found that a reduction of the Setd6 protein level by shRNA mediated knockdown reduces the level of R-spondin 2, being a signaling molecule in the same pathway. Taking this data into account we propose that Setd6 might play a role in signaling, which should be explored in further detail.

Bioinformatic analysis revealed that Setd3 and Setd4 are highly related to Setd6 and that all three proteins have identical domain architectures. This could indicate the involvement in a common pathway or similar regulatory mechanisms. A detailed analysis of those two proteins could help understanding how signaling influences chromatin biology.

In the second part of the thesis we demonstrate how lysine methylation of the non-histone protein p53, a key regulator of the cellular stress response, can alter is activity as a transcription factor. By *in vitro* enzymatic assays we show that the SET-domain containing enzyme Smyd2 is capable of mono-methylating a lysine residue in the C-terminal domain (CTD) of the protein and that the demethylase Lsd1 removes a di-methyl group at the same residue. In cellular assays we demonstrate that both enzymatic activities have an inhibitory effect on p53 mediated transcriptional activation. This regulatory non-histone lysine methylation deepens the understanding of a second layer of epigenetic mechanisms influencing transcription upstream of the chromatin fiber. Future experiments should address how the activity of Smyd2 and Lsd1 are modulated. This would be important giving the fact that p53 is a major tumor suppressor misregulated or mutated in many different types of cancer and thereby offering novel targets for drug development.

Publications:

Huang J, Perez-Burgos L, Placek BJ, Sengupta R, <u>Richter M</u>, Dorsey JA, Kubicek S, Opravil S, Jenuwein T, Berger SL.
Repression of p53 activity by Smyd2-mediated methylation.
Nature. 2006 Nov 30;444(7119):629-32.

Huang J, Sengupta R, Espejo AB, Lee MG, Dorsey JA, <u>Richter M</u>, Opravil S, Shiekhattar R, Bedford MT, Jenuwein T, Berger SL.
p53 is regulated by the lysine demethylase LSD1.
Nature. 2007 Sep 6;449(7158):105-8.

<u>Richter M</u>, De La Rosa-Velazquez IA, Schmidt A, Opravil S, Winter G, Michael E, Jenuwein T
Setd6 is a histone lysine mono-methyltransferase interacting with remodeling activities and components of the Wnt signaling pathway.
In preparation

1.1. Zusammenfassung

Im Bereich der Epigenetik wurden vielen Mechanismen aufgeklärt, die es erlauben zu verstehen, wie es möglich ist das verschiedene Zelltypen eines Organismus zwar das gleiche Genom haben, die Expressionsprofile aber trotzdem unterschiedlich sind. Diese Mechanismen beinhalten die posttranslationale Modifikation der Histone im Chromatin, DNA-Methylierung, trans-aktive nicht-kodierende RNAs und spezialiserte Bereiche im Zellkern, auch „nuclear territories" genannt. Ein wenig erforschter Bereich ist die Möglichkeit eines extrazellulären Signals vom Zytoplasma in den Kern zu gelangen und dort eine Änderung des Transkriptionsprofils herbeizuführen, wie es zum Beispiel bei Differenzierung, Migration oder Apoptose vorkommt. Im ersten Teil dieser Arbeit charakterisieren wir eine neue Histon-Monomethyltransferase (Setd6), die Lysine in Histone H2A, H3 und H4 modifiziert, die bis jetzt nur in ihrer acetylierten Form beschrieben wurden. Von den etwa 50 Methyltransferasen mit einer SET-Domäne wurden verschiedenste für ihre enzymatische Aktivität beschrieben. Die Klasse der Monomethyltransferasen ist dabei stark unterrepräsentiert. Di- und Trimethyltransferasen hingegen wurde mehrfach charakterisiert, wobei manche sogar die gleichen Reste in Histonen modifizieren. Dies könnte die Tatsache reflektieren, dass der monomethylierte Zustand gleichzeitig einen Ausgangszustand darstellt, der vielleicht schon vor der Inkorporation der freien Histone in das Chromatin existiert. Noch interessanter ist, dass die von Setd6 modifizierten Reste identisch mit denen sind, die von dem Histon-Acetyltransferase Komplex Tip60 acetyliert werden. In einer Immunpräzipitation fanden wir Interaktionen zwischen Setd6 und einer Komponente mit Remodelierungsaktivität aus diesem Komplex. Die Daten deuten somit auf eine antagonistische Funktion der beiden Enzyme hin. Tip60 wurde mit der Pluripotenz von embryonalen Stammzellen in Verbindung gebracht und die mögliche Rolle von Setd6 in diesem Bereich wird in zukünftigen Experimenten untersucht werden. Wir haben im selben Experiment β-catenin als Interaktionspartner identifiziert, welches das intrazelluläre Schlüsselmolekül im

Wnt Signalübertragungsweg darstellt. Weiterhin konnten wir zeigen, dass eine durch shRNA vermittelte Verringerung des Setd6 Proteinlevels auch zu einer Verringerung von R-spondin2 (Rspo2) führt, das als Signalmolekül im selben Übertragungsweg fungiert. Diese Ergebnisse lassen auf eine mögliche Funktion von Setd6 in Signalübertragungskaskaden schließen, was in weiteren Studien zu beweisen wäre. Eine bioinformatische Analyse hat gezeigt, das Setd3 und Setd4 von ihrer Aminosäuresequenz und Domänenarchitektur sehr ähnlich zu Setd6 sind. Dies könnte darauf hindeuten, dass sie in einem gemeinsamen regulatorischen Mechanismus agieren. Eine detaillierte Analyse dieser zwei Proteine, könnte helfen zu verstehen, wie Signalübertragungskaskaden die Chromatinstruktur beeinflussen.

Im zweiten Teil der Arbeit zeigen wir wie Lysinmethylierung des nicht-histon Proteins p53, einem Schlüsselprotein in der zellulären Stressantwort, seine Aktivität als Transkriptionsfaktor verändert. Durch enzymatische *in vitro* Experimente können wir zeigen, dass Smyd2, ein SET-Domänen Protein, einen Lysinrest in der C-terminalen Domäne monomethyliert und dass Lsd1 eine Dimethlyierung am selben Rest entfernen kann. In zellulären Experimenten zeigen wir weiterhin, dass beide enzymatischen Aktivitäten eine inhibitorische Wirkung auf die p53 vermittelte transkriptionelle Aktivierung von Genen hat. Diese regulatorische nicht-histon Methylierung hilft dabei eine weitere Möglichkeit der epigenetischen Regulation besser zu verstehen, die die Transkription noch vor dem Chromatinstrang beeinflusst. p53 ist ein wichtiger Tumorsupressor, somit wären weitere Experimente interessant, die die Regulation von Smyd2 und Lsd1 betreffen, da dies helfen könnte neue Ansatzpunkte im Bereich der Tumortherapie zu finden.

2. Introduction

2.1. Epigenetics

Deoxyribonucleic acid (DNA) is the key molecule in any living organism. It allows the production of RNA which is then translated into proteins needed for all cellular processes. The information encoded on the DNA is stably transmitted through cell divisions and the germline. In nearly every cell of a given organism this information is identical thereby posing the question how it is possible that an identical genome can lead to so many cell types that fulfill different and specialized functions. The field of epigenetics investigates the mechanisms responsible for a defined usage of distinct parts of the genome leading to a particular transcriptional profile.

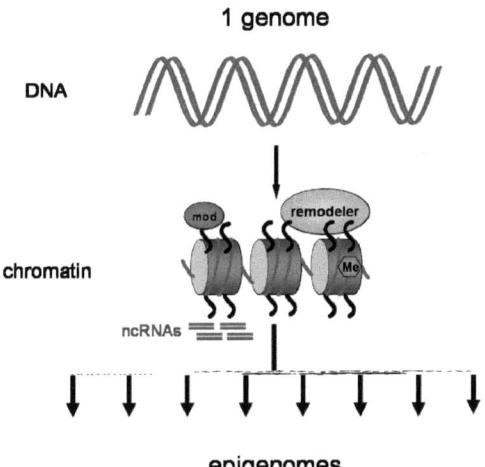

Figure 2-1 By epigenetic mechanisms it is possible that one genome gives rise to many different epigenomes in the various cell types of an organism. The mechanisms include e.g. posttranslational modifications of histones, DNA methylation, nucleosome remodeling and regulatory non-coding RNAs (ncRNAs).

This includes the formation of nucleosomes by wrapping DNA around the four core histones H2A, H2B, H3 and H4, thereby compacting it and allowing and efficient packaging in the nucleus of a cell. This compaction is increased by incorporation of the linker histone H1 and by other nuclear proteins. The histone DNA complex is called chromatin and is the reason why the same genome can give rise to many different epigenomes in different cells as posttranslational modifications (PTMs) such as methylation, acetylation or phosphorylation on the flexible N-terminal part of the core histones can recruit proteins (fig. 2-1) which will e.g. lead to a decrease in the compaction of the chromatin fiber allowing an efficient transcription of the underlying DNA strand.

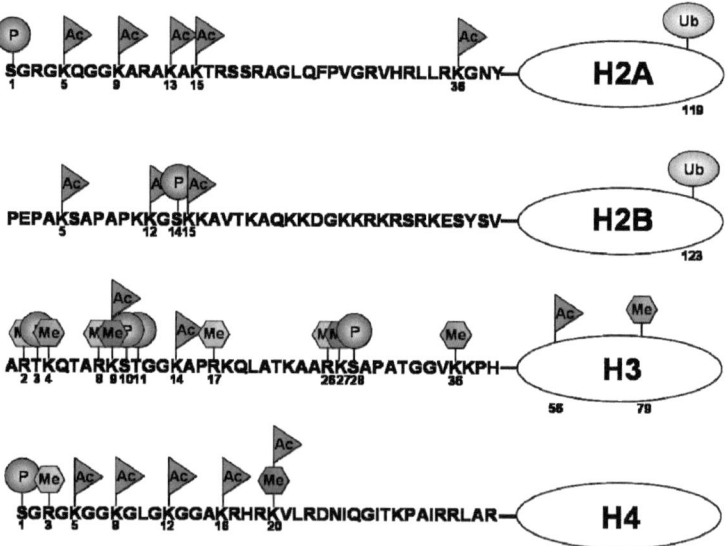

Figure 2-2 Core histones are post-translationally modified by phosphorylation (grey circle), acetylation (blue triangle), arginine methylation (yellow hexagon), ubiquitination (white ellipse) and histone methylation (activating =green hexagon; repressive = red hexagon). Adapted from textbook "Epigenetics".(Allis 2007)

Vice versa, an increased chromatin compaction will prohibit the transcriptional machinery to access it. Other epigenetic mechanisms include the methylation of DNA itself, modifications of the C-terminal part of the core histones often destabilizing the nucleosomal structure, incorporation of specialized histone variants and the recruitment of non coding RNAs (ncRNA). Many residues in the histones have been shown to be posttranslationally modified and their function is at least in part understood (fig. 2-2). As the regulation of genes has to be tightly controlled it is not astounding that the modifications are usually not isolated but can be found in combinations. These combinations allow the binding of specific proteins at different regions of the genome leading to an appropriate transcription of the genes needed for the cell to maintain its identity or to differentiate into a more specialized cell upon a stimulus from its environment. The fact that multiple modifications are needed for these mechanisms led to the proposition of the "histone code" hypothesis which postulates that proteins exist that would recognize these combinations. Indeed proteins have been identified containing multiple domains with one recognizing e.g. an acetylated lysine (bromodomain) while a second would bind to a methylated lysine (PHD finger) on the same or a neighboring histone. A more detailed introduction into epigenetics can be found in the textbook "Epigenetics".(Allis 2007)

2.2.   *Epigenetic regulation by histone lysine methylation*

Histone lysine methylation can be found on several residues in the histone tails and is best characterized for histone H3 and H4. Lysine residues can be mono-, di- or tri-methylated (me1, me2 or me3) (fig. 2-3). Mono-methylation has been described for at least two of the characterized sites to be the initial state and thereby being essential for any following functionality (Schotta et al. 2004; Karachentsev et al. 2005; Huen et al. 2008; Schotta et al. 2008). According to the state of the underlying DNA template, lysine methylation as most of the other modifications on histones is classified as being activating or repressing. The best characterized marks associated with active regions are H3K4me3 and H3K36me3, while H3K9me3, H3K27me3 and H4K20me3 are mainly associated with inactive or silenced regions (Kirmizis et al. 2004; Schotta et al. 2004). Many more methylated lysines are found in the histone tail but a clear classification of those has yet to be established. With the exception of one position (H3K79me) (Steger et al. 2008) all other known positions are modified by histone lysine methyltransferases (HMTases) containing a SET (Su(var)3-9, Enhancer of zeste, Trithorax) domain which catalyzes the transfer of a methyl group from S-adenosyl methionine (SAM) to the substrate.

Figure 2-3 Histone lysine methylation occurs as mono-, di- or tri-methylation. It is catalyzed by histone lysine methyltransferases (HMTs) and can be removed by histone lysine demethylases (HDMs). Monomethylation has been shown for several residues to be the initial state thereby being essential for any following functionality.

In the murine genome approximately 50 genes can be found coding for proteins with such domain. From this list of proteins only some have been characterized to date and the function of the others still has to be explored. H3K4me3 methylation is mainly found at promoters and is indicative for active genes (Schneider et al. 2004; Shilatifard 2008). Two HMTases in *Drosophila melanogaster* (trx and ash1) catalyzing this modification belong to the trithorax group proteins as they are essential for the activity of developmentally regulated genes, such as those in the Hox cluster (Papp and Muller 2006). H3K36 methylation can also be found in the regions of active genes but in the body of the genes and not at their promoters. The molecular function of this modification is yet to be fully understood but initial data indicates that it might be responsible for the repression of alternative transcripts (Carrozza et al. 2005). In yeast the enzyme Set2 catalyzes this methylation and interacts with RNA polymerase II allowing it to travel along the transcribed gene (Li et al. 2003). An indicative mark for inactive or silenced regions of the genome is H3K9me3. In *D. melanogaster* it is catalyzed by the Su(var)3-9 enzyme (Schotta et al. 2002) which has two homologues in the mammalian genome Suv39h1 and Suv39h2. A loss of those enzymes leads to a reactivation of silenced regions in the genome called heterochromatin (Rea et al. 2000). Heterochromatin can be divided into facultative and constitutive. The formation of facultative heterochromatin is often accompanying developmentally regulated transitions from an active to an inactive state as for example polycomb (PcG)-mediated gene silencing (Beisel et al. 2007) or the random inactivation of one of the two X chromosomes in female mammalian cells (Xi) (O'Neill et al. 1999; Vincent-Salomon et al. 2007). Constitutive heterochromatin can be found at intergenic regions mainly composed of repetitive elements and non-coding stretches of DNA. The silenced state is default and the chromatin is compacted through the recruitment of proteins via their chromodomain (e.g. heterochromatin protein 1 (HP1)) by tri-methyl marks at H3K9 and H4K20 (Trojer et al. 2007). The DNA is methylated (Lachner and Jenuwein 2002). H3K27me3 being also a repressive mark is not only important for X inactivation but plays an essential role in gene silencing especially during development and here in concert with the recruitment of proteins from the Polycomb group (PcG). Enhancer-of-zeste and its mammalian homologue Ezh2 are essential genes and catalyze this methylation (Erhardt et al. 2003). Summarizing the given examples we can find HMTases methylating distinct lysine residues in the histone tails thereby creating a binding platform either alone or in concert with other modifications for proteins that lead to compaction or decompaction of the underlying chromatin template. This then ultimately leads to the activation or inactivation of the genomic region. For a long time histone lysine methylation and here particularly the tri-methyl state was thought to be a stable modification as no enzymes were known that could reverse this reaction. But by now several groups have shown that a class of hydroxylases containing a so called JmjC domain can demethylate tri-methylated lysine residues (Klose et al. 2006; Takeuchi et al. 2006; Whetstine et al. 2006). This is in analogy to any of the other histone modifications as also for acetylation deacetylases exist or for the phosphorylation several phosphatases are described.

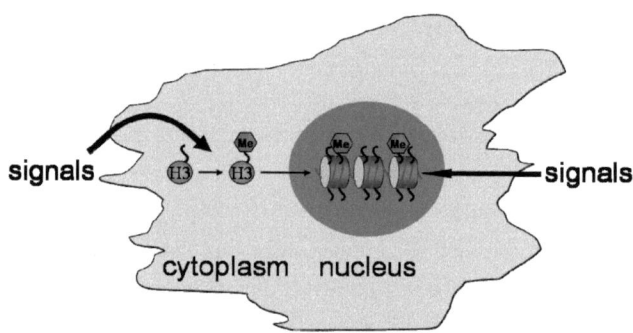

Figure 2-4 Signals from the cellular environment cause changes in the transcriptional profile by alterations of the chromatin fiber by e.g. histone lysine methylation. It has been shown that the modifications on the histones can occur in the cytoplasm prior to the incorporation into the nucleosome or after it.

An interesting question for epigenetic regulation is how extracellular signals are translated into alterations of the chromatin fiber leading to changes in the transcriptional profile of a given cell type. One possibility is the activation of histone modifying enzymes in the nucleus which are targeted towards specific loci either by their other domains or by additional proteins. The second possibility is that the pool of free histones is modified either in the cytoplasm or the nucleus prior to the incorporation into nucleosomes (fig. 2-4) (Loyola et al. 2006). This could induce a broader alteration of the expression profile as e.g. in the case of cellular senescence, where the transcription of many genes is affected (Imai and Kitano 1998). The examples given above are to day the best characterized methyl-marks according to their biological function. Many more sites for methylation have been found by mass spectrometry on the histone tails and cores but their function has to be elucidated. In the work presented here we have used a candidate screen using recombinant SET-domain containing proteins with unknown function and recombinant histone octamers as substrates in a radioactive HMTase assay to find novel interesting activities which where then followed up in a more detailed analysis.

2.3. Interplay between histone modifications

Histone modifications are involved in various cellular processes that reach from proliferation over the defined regulation of gene transcription in response to e.g. the activation of a signaling pathway to apoptosis (Bonisch et al. 2008). Considering gene regulation, interplay means that certain marks will have a positive effect on each other reinforcing the original signal while others will have opposing functions as they e.g. differentially regulate genes involved in several pathways (Suganuma and Workman 2008). In general histone lysine acetylation has been often linked to gene activation while methylation of histones and DNA is more often found in inactive regions of the genome. Genome-wide ChIP studies together with the analysis of mutant mice have allowed a first understanding of the cascades of gene activation or general gene silencing in part. Ubiquitination of lysine 123 in H2B leads to tri-methylation of H3K4 during the activation of a gene (Shukla et al. 2006) which then leads to the acetylation of at least H3K9 and H4K16 (fig.2-5a) (Berger 2007; Latham and Dent 2007). Those acetyl marks recruit bromodomain containing remodeling complexes needed to open the chromatin structure to allow binding and progression of transcription factors and the RNA polymerase II machinery. For the formation of heterochromatin the first histone mark to be set is H3K9me3 by the Su(var) enzyme, which is followed by the binding of HP1 recruiting the Suv4-20 enzymes that will tri-methylate H4K20 (Schotta et al. 2004) and recruit deacetylases. This will ultimately lead to DNA methylation and the compaction of chromatin most likely via the binding of compacting proteins (fig.2-5b) (Latham and Dent 2007). Also other interplays are known that are more limited to certain classes of genes as for example H3S10 phosphorylation (H3S10ph) in concert with H3K14 acetylation leading to the activation of genes in response to beta interferon activation (Walter et al. 2008). H3S10ph is also very important for cell cycle dependent condensation of chromosomes during mitosis in combination with H3S28ph and H3T11ph (Nowak and Corces 2004). In addition also special events in the cell can trigger cascades of chromatin modifications such as DNA double strand breaks (DSB).

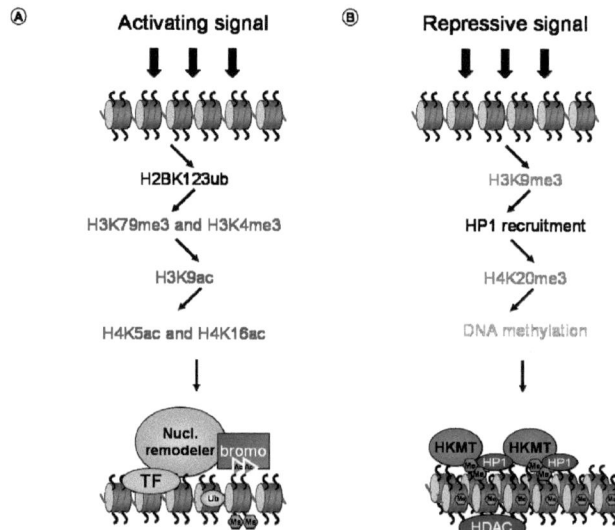

Figure 2-5 To allow a specific transcriptional read-out combinations of histone modifications are used allowing a stringent control and multiple layers of regulation. Adapted from textbook "Epigenetics".(Allis 2007)

At breakpoints the phosphorylation of S129 in H2AX is induced by the activated ATM kinase which then leads to the recruitment of cohesion to stabilize the broken ends (Watrin and Peters 2006; Iijima et al. 2008). Furthermore the Tip60 acetyltransferase complex is recruited leading to the acetylation of histones H3 and H4 (Squatrito et al. 2006). H4K20 gets methylated by the Suv4-20 enzymes and this methylation is recognized and bound by the 53BP1 protein inducing non-homologous end-joining to repair the DSB (Kobayashi et al. 2008). Interestingly in response to DNA-damage H3K4me3 at promoters of genes important for cell cycle progression recruits the repressor Ing2 and promotes deacetylation and inactivation (Shi et al. 2006). Those few examples show two interesting principles of histone modifications. One is that their function depends largely on the context where they are found and this has to be taken into account when making a prediction for the induced alteration.

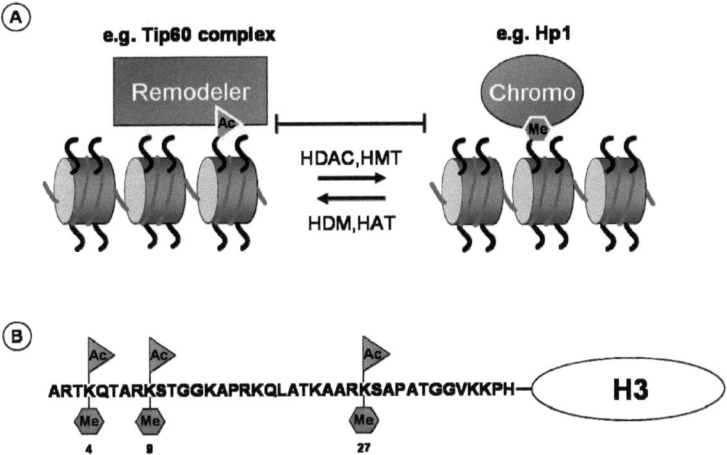

Figure 2-6 a) Acetylation and methylation can have opposing function either by the recruitment of proteins with different functions (compacting vs. decompacting) or b) by targeting the same residue and thereby being mutually exclusive. For H3 three of such excluding modifications are described.

The second is that modifications influence each other most likely via the recruitment of other factors (fig. 2-6 a) but can also directly exclude each other by targeting the same residue (fig. 2-6b). The best example here is the methylated (silencing) or acetylated (activating) form of H3K9. H3K9ac prevents in the region of active genes the methylation of H3K9 which would lead to the recruitment of HP1 and terminally to silencing. In this thesis data is presented that indicates the existence of new such pairs of acetylated or methylated residues which could have an antagonistic role in gene regulation.

2.4. Non-histone lysine methylation

Histones and particularly their N-terminal parts are amongst the most heavily posttranslationally modified peptides in the proteome. They are targeted by a plethora of modifying enzymes and the modifications have diverse functions like gene regulation, chromosome stability, cell cycle progression or silencing of large domains in the genome. However, in addition to histones, many other proteins are phosphorylated, acetylated, methylated etc. Especially phosphorylation cascades have been well characterized in the last decades as documented in many publications on e.g. the mitogen-activated protein kinase (MAP-kinase) pathways. It has been demonstrated that misregulation of phosphorylation events can strongly affect cellular processes often leading to the development of cancer (Lawrence et al. 2008). Acetylation of non-histone proteins is also a well known PTM. The most prominent example for this PTM is the tumor suppressor p53 which is the gene with the highest mutation rate over all cancer types. It plays a crucial role in the regulation of cell cycle progression and apoptosis after stress induced damage (Bode and Dong 2004; Toledo and Wahl 2006). Having such an important role, its function has to be tightly regulated and the most obvious way are PTMs as they allow the crosstalk with other regulators or pathways. The C-terminal domain is also known as the regulatory domain as it is targeted by many modifying factors (Bode and Dong 2004). Its amino acid sequence is very similar compared to the N-terminal tail of the core histones H3 and H4 containing multiple lysine, serine, threonine and arginine residues that are or can potentially be modified. One of the modifying factors is the acetyltransferase CBP/p300 which targets several lysine residues in this domain and thereby increases the transcription-factor activity of the molecule (fig.2-7b) (Bode and Dong 2004). The modifications and activation are reversed by the deacetylase HDAC1. The unmodified peptide is prone to ubiquitination via Mdm2 which leads to the degradation of p53 (fig.2-7a) (Toledo and Wahl 2006). An example where acetylation leads to a decreased DNA binding affinity is the transcription factor Yin Yang 1 (YY1), which, depending on the context and binding partners, can have an activating or repressive function and is important for differentiation and development. Two sites can be acetylated by CBP/p300 and reversed by class 1 HDACs (HDAC1, HDAC2 and HDAC3). The first site, when acetylated, leads to a lower DNA binding affinity and the second site, when acetylated, leads to an altered transcriptional activity (Glozak et al. 2005). But not only transcription factors are targets for non-histone lysine acetylation. Also the nuclear receptors like the androgen receptor (AR), the estrogen receptor α or the orphan receptor SHP (short heterodimer partner) are targets and several other unrelated proteins including viral proteins potentially playing a role in cellular defense mechanism.(Glozak et al. 2005) Non-histone lysine methylation is a less explored field but given the number of SET-domain containing enzymes in the genome it is very likely that additional targets and thereby functionalities will be reported. As for acetylation the best studied protein for this PTM is p53 as there is great interest to understand its regulation for the development of anti-cancer drugs. While acetylation leads to an increase of its activity as a transcription factor the effect of methylation depends on the position of the modified lysine and the degree of methylation (mono-, di- or tri-methylation). SET-domain-containing protein-9 (Set9), originally shown to mono-methylate H3K4, was the first enzyme to be shown to mono-methylate p53 at lysine 372 leading to its stabilization and activation (fig.2-7b) (Chuikov et al. 2004). It also recruits TIP60 which will acetylate p53 increasing its activity (Squatrito et al. 2006).

A) p53 in unstressed cell:

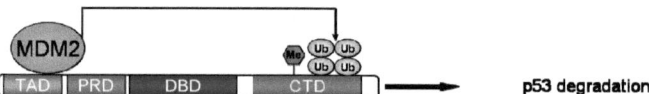

p53 degradation

B) p53 after stress (e.g. DNA damage):

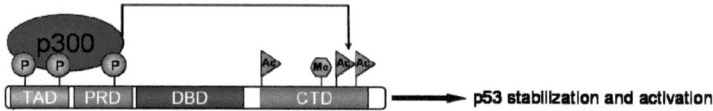

p53 stabilization and activation

Figure 2-7 a) p53 is ubiquitinated in unstressed cells by MDM2 leading to its degradation keeping the pool of free molecules constant. By repressive lysine methylation the activity of the molecule is reduced offering a second level of control. b) In stressed cells p53 is acetylated and methylated leading to an increased activity.

By contrast, a mono-methylation on lysine 382 in p53 has a repressive function (fig.2-7a) and the protein level of PR-SET7, the catalyzing methyltransferase, which is already well characterized for its mono-methylating activity towards H4K20, is decreased following DNA-damage.(Shi et al. 2007) Other non-histone targets that are methylated by SET-domain containing proteins are Transcription initiation factor TFIID subunit 10 (TAF10; increased affinity for PolII), ERα (activation), Vascular endothelial growth factor receptor 1 precursor (VEGFR1, enhanced activity) and Euchromatic histone-lysine N-methyltransferase 2 (allows binding to HP1α and HP1γ) (Chin et al. 2007; Huang and Berger 2008; Rathert et al. 2008). In this thesis, SET and MYND domain-containing protein-2 (Smyd2) is presented to mono-methylate the tumor suppressor p53 at lysine 370 and thereby repressing its function. Furthermore we show that the activating di-methylation of lysine 370 for which the HMT is unknown can be removed by the demethylase LSD1.

3. Setd6 is a histone lysine mono-methyltransferase interacting with remodeling activities and components of the Wnt signaling pathway.

Richter M, De La Rosa-Velazquez IA, Schmidt, A, Opravil S, Winter G, Michael E, Jenuwein T

3.1. Abstract

In this work we present the SET-domain-containing protein Setd6 as a novel histone lysine mono-methyltransferase modifying histone H1K26, H2AK5, H3K14, H4K5 and H4K12. We show that the gene is ubiquitously transcribed throughout murine embryonic development and also in all tissues examined which could be explained by a general role in chromatin biology. Intriguingly, the modified residues have been described in their acetylated form only. By reducing the protein level of Setd6 in NIH 3T3 cells using a shRNA mediated knockdown, we find an alteration in the transcriptional profile using micro array analysis. This alteration is not genome-wide but rather restricted to a defined set of genes. The most down regulated gene is R-spondin 2 (Rspo2) being a signaling molecule in the Wnt pathway possibly implying a role of Setd6 in its regulation. This notion is strengthened by the fact that by immunoprecipitation from murine embryonic fibroblasts (MEFs) we identify β-catenin as a specific interactor. This molecule is the key regulator of the intracellular response to Wnt signaling. Another interacting protein is Tip49a being a DNA/RNA helicase which is a subunit of different remodeling complexes but its function has been described in detail for the Tip60 complex. The Tip60 protein is a histone acetyltransferase capable of modifying the same residues as Setd6 on the histone tails. Furthermore, the complex has been shown to be essential for the maintenance of pluripotency in embryonic stem cells. We propose that Setd6 has an antagonistic role to this complex by targeting the same residues and by its connection to the Wnt pathway which has also been demonstrated to be important for pluripotency and defined differentiation throughout murine development.

3.2. *Introduction*

3.2.1. SET domain containing enzymes

The SET domain was first described in 1994 and was found to be an evolutionary conserved sequence motif that is present in many proteins that are important for the regulation of the transcriptional state of genomic regions (e.g. Polycomb- and trithorax group) (Tschiersch et al. 1994). One protein that was characterized in greater detail was Su(var)3-9 in *Drosophila melanogaster* being a dominant modifier of position-effect variegation (PEV) and its mammalian homologue Suv39H1 which was shown to affect chromosome segregation and mitotic progression indicating a role in higher order chromatin structure (Jenuwein et al. 1998; Aagaard et al. 1999). The molecular mechanism was only understood after the discovery that the SET-domain has an enzymatic activity allowing it to transfer a methyl-group from its cofactor SAM to a lysine residue of a substrate protein. (Rea et al. 2000) For Suv39H1 the substrate is histone H3 at lysine 9 and this modification since then has been demonstrated to be important for the silencing of repetitive elements, for proper mitotic progression due to a crosstalk with H3 serine 10 phosphorylation (H3S10ph) and for modulating the transcriptional activity of genes.(Jenuwein 2001). In the murine genome approximately 50 SET domain containing proteins are encoded and can be classified according to it. This classification can be visualized in a phylogenetic tree (fig. 3-1). From this collection of enzymes roughly half have an assigned function and their target residues are known; the other half has yet to be explored. The list of proteins can be divided in two subfamilies. One subfamily consists of the enzymes with a canonical SET-domain while the other has a so called PR-domain (PRDI-BF1-RIZ1 homologous region), which is highly similar to the SET-domain and has been shown to also be catalytically active. Therefore it is thought to be an evolutionary derivative of the first class (Huang et al. 1998; Hayashi and Matsui 2006). A prominent member of this family is Pr-set7, mono-methylating

H4K20, being one of the few mono-methyltransferases published. Its activity is important for murine development (Karachentsev et al. 2005; Huen et al. 2008). Riz1, also a member of this family, has been shown to interact with the retinoblastoma tumor suppressor protein (Rb) and to be important for a cell cycle arrest and the induction of apoptosis (He et al. 1998). Its transcript number is reduced in breast cancer and it acts as a transcriptional repressor by methylation of H3K9 (Xie et al. 1997; Kim et al. 2003). In addition to the active domain many of the enzymes have two other conserved motifs called the pre- and post-SET domain, with the first being important for the structural integrity of the protein and the second for substrate binding. Several SET domain-containing proteins are large multidomain proteins often containing protein-protein interaction domains, DNA-binding domains, or domains for the binding to posttranslationally modified histones. This multidomain architecture is needed to tightly control their binding and activity in the genome, which at the same time explains why many of them target the same residues in the histone tail and modify them to the same state, but still have a different impact on chromatin structure and transcriptional regulation. The characterization of novel activities from those enzymes is needed to elucidate epigenetic regulation in greater detail as we still do not understand many of the regulatory circuits in the genome. For example the histone code hypothesis postulates that the combination of histone marks defines the transcriptional and structural state of the underlying chromatin template (Jenuwein and Allis 2001). To prove this idea we have to understand which enzymes are active in defined regions of the genome, how the cross-talk between different enzymes and marks is established and whether certain patterns can be linked to defined functions (Sims and Reinberg 2008). In this thesis we describe a novel enzyme (Setd6) that belongs to the class of mono-methyltransferases in the PR-domain containing family and modifies residues that have so far been mainly characterized in their acetylated form. Those residues have been associated with active transcription, cell cycle progression and histone deposition. As methylation and acetylation are mutually exclusive we propose an antagonistic role. The fact that this enzyme is a mono-methyltransferase is important giving the fact that such activity is underrepresented in the catalogue of described enzymes from the complete list of SET-domain-containing proteins and it has been shown that for histone H4 lysine 20 and histone H3 lysine 9 mono-methylation is the initial state and thereby makes the enzymatic activity essential.

3.3. Results

3.3.1. Setd3, Setd4 and Setd6 have a common domain architecture

Using the phylogenetic SET-domain tree (fig. 3-1) and published data on methyltransferases we identified candidate genes for activity assays searching for novel mono-methylating enzymes.

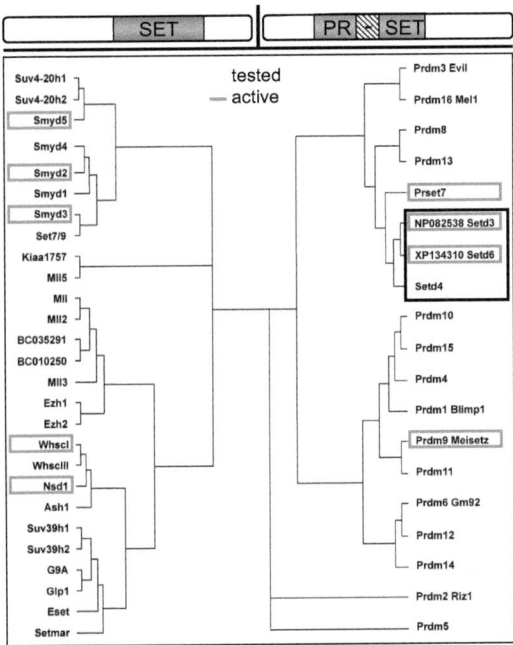

Figure 3-1 SET-domain containing enzymes were analyzed for their similarity in the SET-domain. Proteins on the left contain the canonical SET domain and those on the right a PR-SET domain. Yellow and red boxes indicate enzymes that have been tested for their methylating activity towards recombinant histones (yellow=inactive; red=active).

We chose 22 enzymes for which no activity was described and which are similar in their SET-domain to proteins with an important function, of those 9 showed activity in a radioactive histone methyltransferase assay using the recombinant enzymes and recombinant free histones as substrate (fig. 3-1). One subfamily from the PR domain-containing part of the tree was of special interest as its enzymes Setd3, Setd4 and Setd6 are highly similar in their SET domain to Prset7, being one of the few published mono-methyltransferases and being essential for murine development (Karachentsev et al. 2005; Huen et al. 2008). Setd6 showed robust activity in our assay. Bioinformatic analysis revealed that the 3 enzymes are not only highly related in their methyltransferase domain but also in their general domain architecture (fig.3-2).

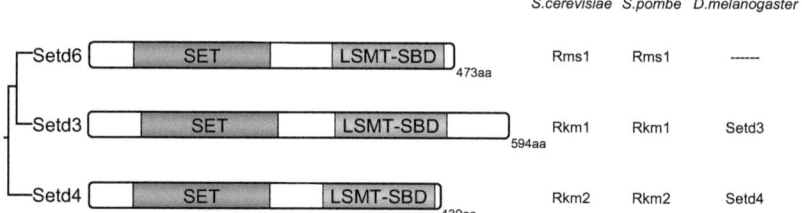

Figure 3-2 Setd3, Setd4 and Setd6 share a common domain architecture. In addition to the N-terminal SET domain they have a C-terminal LSMT-SBD which could mediate the binding to their substrates.

All three have an N-terminal SET-domain and a C-terminal, so called Rubisco LSMT substrate-binding domain, first described in the RuBisCO large subunit methyltransferase (LSMT), where it is important for the binding to the N-termini of histone H3, H4 and the large subunit of the Ribulose-1,5-bisphosphate carboxylase/oxygenase (RuBisCO) (Trievel et al. 2003). This domain makes the proteins unique amongst the murine SET domain-containing enzymes and could indicate the involvement in a common pathway and could hint to a function in metabolic regulation. The murine proteins have so far not been examined while in *S.cerevisiae* their orthologs have been partially characterized. The orthologue of Setd3 is Rkm1 (ribosomal lysine (K) methyltransferase 1) and has been shown to di-methylate the ribosomal large subunit protein L23a (Rpl23ab) indicating a role in translational regulation (Porras-Yakushi et al. 2005). The orthologue of Setd4, Rkm2 methylates a large subunit of the ribosome (Rpl12ab) to a tri-methylated state (Porras-Yakushi et al. 2006). Rms1 is the orthologue of Setd6 and has only recently been show to mono-methylate the ribosomal protein Rpl42ab (Webb et al. 2008) which is in contrast to its nuclear localization published previously (Huh et al. 2003).

3.3.2. Setd6 localizes to the nucleus in murine NIH 3T3 cells

To get a better insight into a putative biological function of Setd6 and Setd3 we raised rabbit polyclonal antibodies against the full-length recombinant protein. Those antibodies were tested on western blots against the recombinant protein and whole cell extracts of murine NIH 3T3 showing specific signals at the correct molecular weight of the two proteins (fig.3-3a and b right panels). Setd6 consists of 473 amino acids and has an expected molecular weight of 53kDa, while Setd3 consists of 594 amino acids and the expected molecular weight is 67kDa. For the initial characterization we were interested in the subcellular localization of the enzyme to allow insight into its involvement in epigenetic regulation, as a protein involved in such processes would likely be localized to the nucleus and probably bound to chromatin. And indeed in mouse NIH 3T3 cells Setd6 is localized to the nucleus in a dot-like fashion excluding nucleoli and pericentric heterochromatin with a weak staining in the cytoplasm (fig. 3-3a).

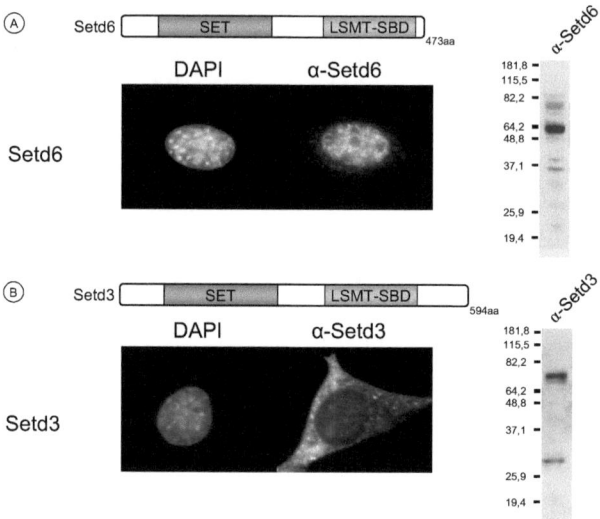

Figure 3-3 a) Setd6 localizes to the nucleus in murine NIH 3T3 cells, while Setd3 is enriched in the cytoplasm (b). On the right western blots with NIH 3T3 whole cell extracts are shown. Expected sizes: Setd6 53kDa; Setd3 67kDa. Antibodies used: Setd6 #1456; Setd3 #1457

With this result we can postulate an involvement in chromatin biology. We also raised antibodies against recombinant Setd3 and in contrast to Setd6 a pronounced cytoplasmic staining is observed with very little signal in the nucleus (fig. 3-3b), which could indicate that the enzyme methylates cytosolic non-histone substrates in analogy to the yeast protein.

3.3.3. Setd6 is transcribed in adult tissues and throughout mouse development

To address the expression profile of Setd6 we asked whether its function would be limited to a certain period in mouse embryonic development or to specific tissues in the adult mouse. For this purpose we used commercially available northern blots (BLOT3 (Sigma Aldrich), MTNI and II (Clontech)) that contain equal amounts of total RNA either from different stages of the developing embryo or from different tissues. As a probe we used a 1kb PCR amplicon of the 1.6kb coding region of Setd6 from the plasmid pDONR/Zeo/Setd6 (primer: Setd6 RT primer1F and 3R) (fig. 3-4) which we labeled using a random primer strategy with the Klenow-fragment and [α-32P]dATP. This analysis revealed that the Setd6 transcript is present at equal levels throughout all probed stages of mouse development and can be found in all adult tissues with high levels in the murine liver, prostate, kidney, salivary gland, heart and thyroid (fig. 3-4). The enzyme although ubiquitously transcribed seems to have important roles for specific cell types in certain organs. For Setd3, using the complete coding region from plasmid pDONR/Zeo/Setd3 (primer: Setd3north forward and reverse) as probe we obtained a similar result but in addition the transcript level was higher in testis and brain (fig.3-5).

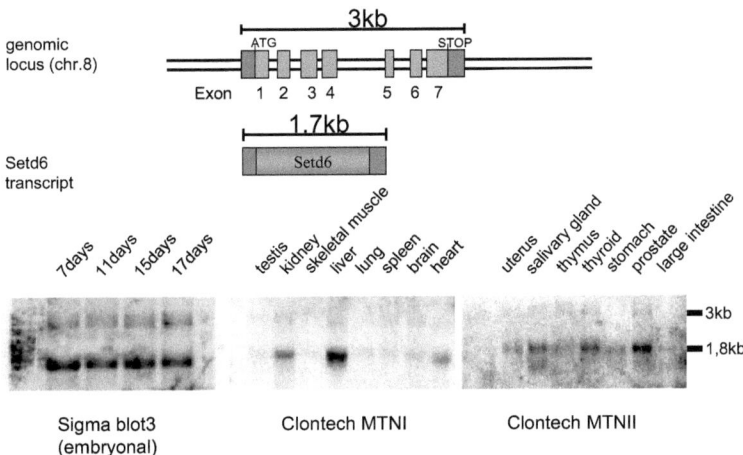

Figure 3-4 The Setd6 genomic locus consists of 7 exons and encodes a 1.7kb transcript, which can be detected throughout murine embryonic development and in some adult murine tissues but at different levels.

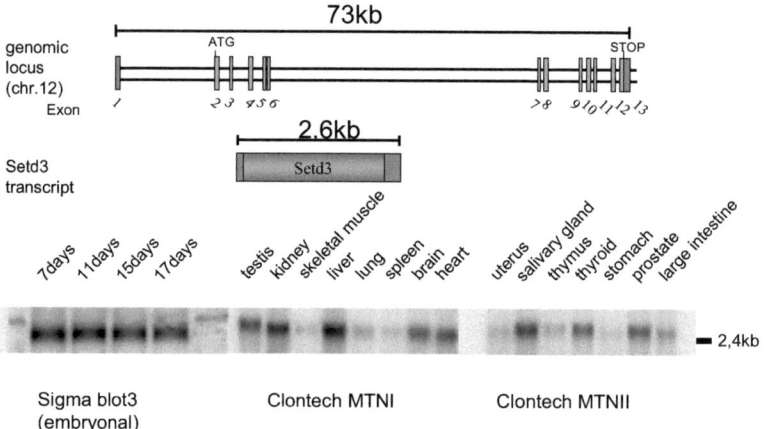

Figure 3-5 a) The Setd3 genomic locus consists of 13 exons and encodes a 2.6kb transcript, which can be detected throughout murine embryonic development and **b)** in all adult murine tissues but at different levels.

3.3.4. Setd6 methylates histone H1, H2A, H3 and H4 in vitro

As we were interested to find novel factors involved in epigenetic regulation we investigated whether murine Setd6 or Setd3 are capable of methylating histones by performing radioactive histone lysine methyltransferase assays (HMTase assays) using the full-length recombinant protein expressed in *E.coli*. The enzyme was incubated with tritium-labeled S-adenosyl methionine (SAM) and one of the core histones or the linker histone H1 (H1.2). An inactive point mutant (Setd6mut; Y285A) replacing a conserved tyrosine in the SAM binding pocket served as a control for Setd6 activity thereby linking the activity to the SET domain. The reactions were run on a 15% SDS-PAGE to separate enzyme substrate and cofactor and blotted to a Polyvinylidene Difluoride (PVDF) membrane before the exposure to an autoradiography film. In this assay Setd6 exhibited broad substrate recognition with activity towards histone H1, H2A, H3 and H4 while H2B was methylated to a lesser extent as can be seen by an overlay of the film and membrane (fig. 3-6).

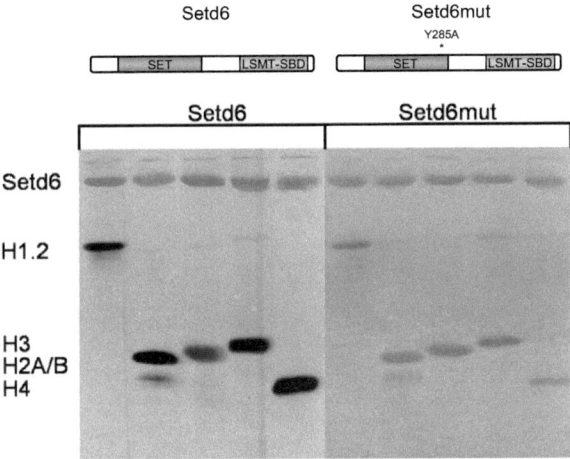

Figure 3-6 Setd6 methylates recombinant histone H1.2, H2A, H3 and H4. An inactivating point mutation in the SET domain (Y285A) abolishes the activity. Setd6 has very little activity towards histone H2B

Such a broad substrate specificity as for Setd6 was so far not reported for any SET domain-containing enzyme and led us to a more detailed analysis of this enzyme. In a similar HMTase assay Setd3 showed very low activity towards histone H3 and no activity towards histones H1, H2A, H2B and H4.

3.3.5. Setd6 is a histone lysine mono-methyltransferase

To be able to address a function of Setd6 in the regulation of chromatin structure or transcriptional regulation we asked which residues are targeted by Setd6 using a peptide based HMTase assay coupled to a mass spectrometric (MS) analysis (MALDI-TOF) allowing us to determine the state of methylation (mono, di or tri) and to identify the modified residue. The peptides in this reaction spanned the N-terminal tails of the histones containing most of the modified amino acids described in the literature. Setd6 was incubated with SAM and with one of the peptides overnight to produce a high percentage of the methylated product for the following MS measurements. The methylated peptides were subjected to a MS/MS analysis, which by fragmentation of the peptide provides the information necessary to attribute the methylation to a certain amino acid.

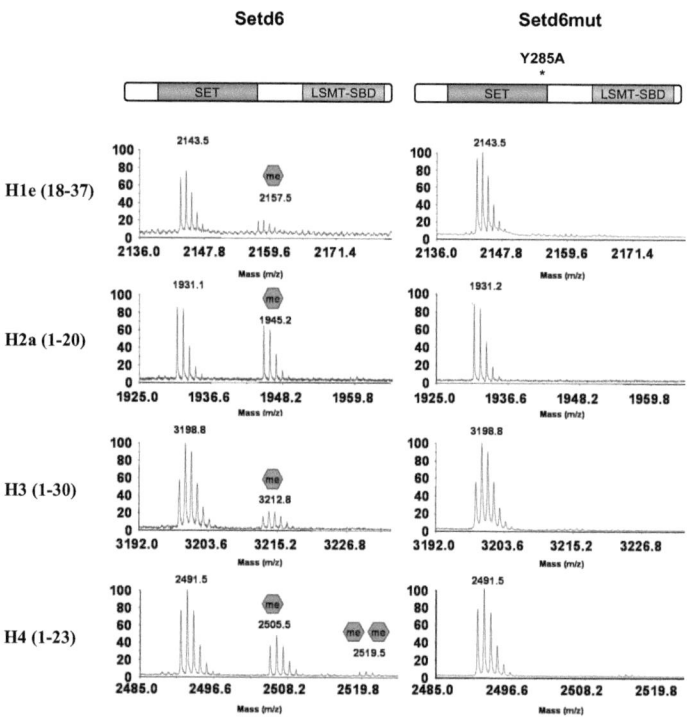

Figure 3-7 In vitro HMTase assay with recombinant Setd6 on N-terminal histone peptides. In the reaction products an additional peak at +14Da indicating a mono-methylation is detected. The H4 peptide also displayed a peak for a doubly methylated peptide (left lower panel). No peaks are present in the spectra of the peptides that were incubated with the inactive Setd6 point mutant (Y285A) (right panel).

The spectra obtained from these measurements clearly showed that Setd6 mono-methylates lysine 5 in H2A, lysine 14 in H3 and lysine 26 in histone H1 indicated by a mass-shift of 14Da equaling the mass of one additional methyl group. In the spectrum of the H4 peptide also a weak signal for di-methylation (+28Da) was detected which would be in contrast with the mono-methylation found on the other peptides (fig.3-7). This additional signal was resolved by the MS/MS analysis showing that in the N-terminal tail of H4 two residues are methylated by Setd6 (K5 and K12). The activity found in those assays could clearly be attributed to the enzymatic activity as the inactive Setd6 point mutant did not produce any methylated product.

3.3.6. Setd6 methylates H1K26, H2AK5, H3K14, H4K5 and H4K12

The mass spectrometric data indicate that Setd6 targets H1K26, H2AK5, H3K14, H4K5 and H4K12. Since we used peptides spanning only the N-termini of the histones we can not exclude that other methylation sites could be present in the histone cores. To clarify this, we produced full length recombinant histones in *E.coli* and introduced point mutations (lysine to alanine) for the identified sites. In a radioactive HMTase assay using tritium labeled SAM we tested whether the signal for methylation disappears when the lysines are replaced for alanines. If the signal would only be reduced additional sites are present in the core. For H1, H2A and H3 the methylation is clearly lost when the point mutations are introduced (fig.3-8 a). The signal for H4 is only slightly impaired when K5 is mutated and strongly reduced for the K12 mutant. The signal is lost only when both sites are mutated (fig.3-8 a).

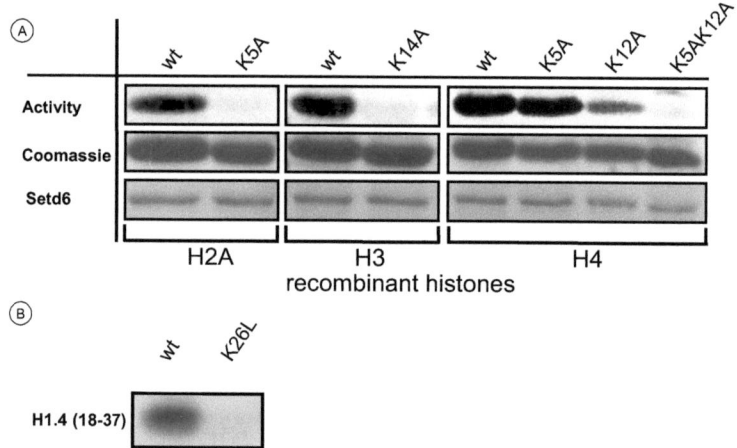

Figure 3-8 Histones mutated at the indicated lysines demonstrate that no additional residues in the histone cores are targeted by Setd6. For H4 both identified sites have to be mutated to abolish the activity entirely. b) For histone H1.4 a mutation of lysine 26 in a peptide spanning amino acids 18-37 abolishes the activity of Setd6

For histone H1.4 we tested a wildtype and a K26L point mutated peptide spanning amino acids 18-37, as we were not able to produce the full-length recombinant histone (fig.3-8 b). Combining the mass spectrometry results and the data on the mutant histones we can conclude that Setd6 is a novel histone mono-methyltransferase targeting sites for which methylation was found previously using mass spectrometry and bulk histones extracted from cells (fig. 3-9) (Kubicek S., Jenuwein T. unpublished data).

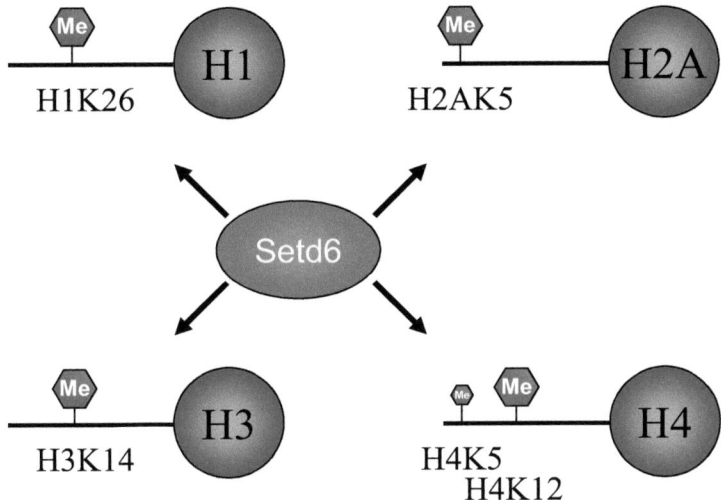

Figure 3-9 Setd6 mono-methylates *in vitro* H1K26, H2AK5, H3K14, and H4K12. H4K5 is a minor target based on the mass spectrometry data and HMTase assay on mutant recombinant histones.

The enzyme responsible for this methylation was unknown and also the function it may have in chromatin biology is unexplored. On the other hand, acetylation has been described in the literature for those lysine residues being involved in transcriptional regulation (H3K14; H2AK5) (Schiltz et al. 1999; Cuddapah et al. 2009), histone deposition (H4K5; H4K12) (Turner 2000) or mitotic progression (H3K14) (Walter et al. 2008). An antagonistic role for Setd6 can be proposed at this point of the analysis.

3.3.7. Reduced Setd6 protein level in cells leads to decreased methylation of Setd6 target residues

The activity of Setd6 was to this point based on *in vitro* assays. Reasoning that a reduction of the enzyme level in NIH 3T3 cells should also lead to a decrease of methylation at the Setd6 target residues, we used a lentiviral based approach to knockdown Setd6 with short hairpin RNAs (shRNAs) specific to its mRNA (fig. 3-10). We tested twelve different shRNAs of which two (shRNA 3 and 13) caused a pronounced decrease as measured by real time PCR (data not shown) and western blot 96 hours post-transduction (fig.3-10).

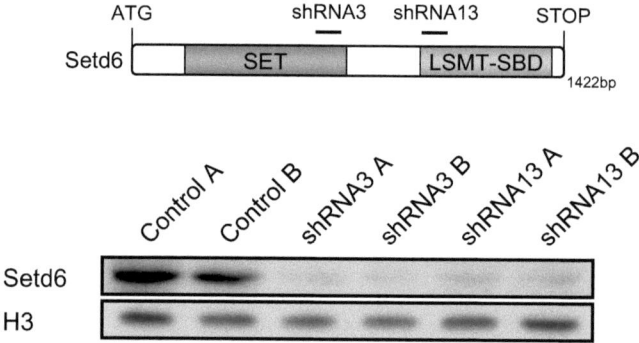

Figure 3-10 The Setd6 protein level can be efficiently reduced by shRNA mediated knockdown in murine NIH 3T3 cells 96 hours after viral transduction (A and B represent two independent experiments). On top the sites in the Setd6 transcript targeted by the two independent shRNAs are depicted. Antibody used: Setd6 #1452

To be able to analyze the *in vivo* function of Setd6 in detail we raised rabbit polyclonal antibodies specific to the methylated target residues. We used branched peptides coupled to Keyhole limpet hemocyanin (KLH) according to (Perez-Burgos et al. 2004). Per epitope three rabbits were used to derive antisera. From those we selected the most specific one according to a standardized dot-blot analysis and quality criteria established by Susanne Opravil in the laboratory of Thomas Jenuwein.

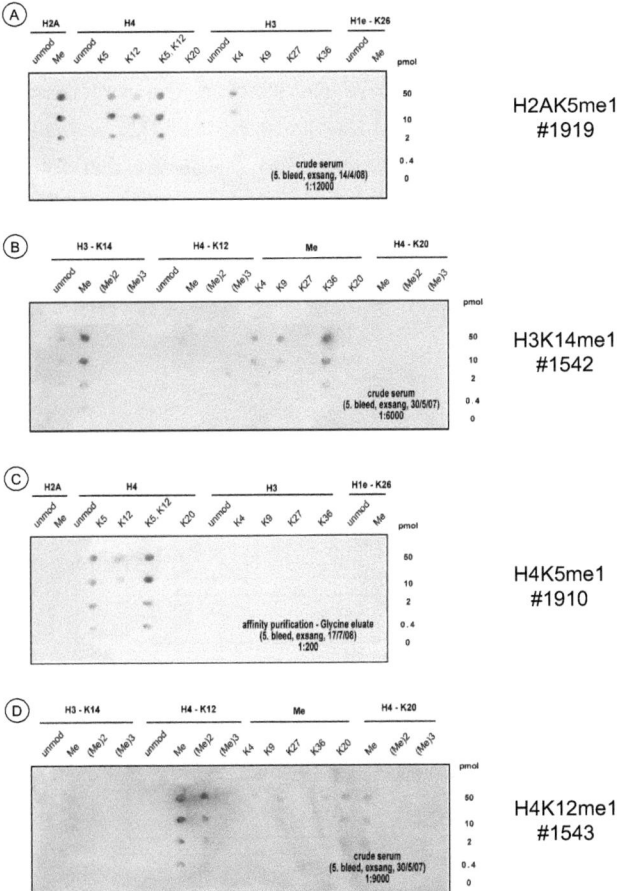

Figure 3-11 a) H2AK5me1 antiserum #1919, b) H3K14me1 antiserum #1542, c) H4K5me1 antiserum # 1910 (affinity purified) and d) H4K12me1 antiserum #1543 displayed best specificity in the dot-blot quality control and were used for all subsequent analyses.

We chose the following sera for all subsequent experiments: H2AK5 me1 antiserum #1919, H3K14me1 antiserum # 1542, H4K5me1 antiserum # 1910 (affinity purified) and H4K12me1 antiserum #1543. Those sera displayed lowest cross reactivity with other methyl-marks (fig. 3-11 a to d). With those reagents at hand we now asked whether the *in vitro* enzymatic activity of Setd6 reflects the *in vivo* situation. For this we infected cells with shRNA 3 and a non-silencing control and made whole cell extracts 4 days after infection. The extracts were analyzed by western blot using the methyl-specific antibodies. This analysis revealed a clear reduction in the level of H2AK5me1, H4K5me1 and H4K12me1, for H3K14me1 the signal is only slightly reduced (fig. 3-12). We also used an antibody specific to H1K26me1 but from the results no argument can be made, most likely due to cross-reactivity with H1K26me2 and H1K26me3 (data not shown).

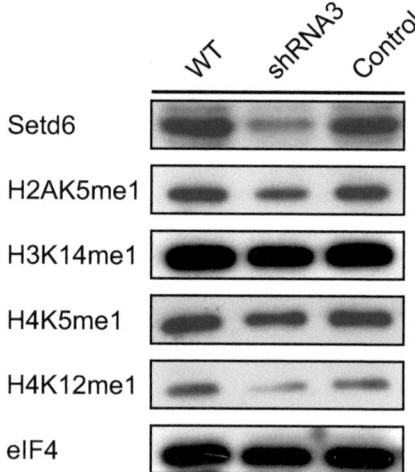

Figure 3-12 A reduction of the Setd6 protein level (middle lane) leads to a reduction of H2AK5me1, H4K5me1 and H4K12me1. H3K14me1 is only slightly affected. α-eIF4 was used as a loading control. Whole cell extracts were prepared 96 hours after viral transduction of NIH 3T3 cells. Antibodies used: Setd6 #1452; H2AK5me1 #1919; H3K14me1 #1542; H4K5me1 #1910; H4K12me1 #1543; eIF4 #2013s (Cell Signaling)

The data indicate that a reduction in the level of Setd6 is linked to a lower amount of histones methylated at the identified sites. Thus, the *in vitro* assays are to a large degree consistent with the enzymatic activity of Setd6 *in vivo*. In addition Setd6 is a SET domain-containing mono-HMT targeting more than one histone. Such a broad specificity was so far only observed for histone acetyltransferases (Berger 1999). Importantly the acetyltransferase Tip60 acetylates the same residues *in vitro* and *in vivo*.(Kimura and Horikoshi 1998; Squatrito et al. 2006)

3.3.8. Staining patterns for H1K26, H2AK5, H3K14, H4K5 and H4K12

To understand the *in vivo* role of Setd6 mediated methylation we used our polyclonal rabbit antibodies specific to the enzyme and the associated methyl-marks for immunofluorescence staining in NIH 3T3 cells comparing their localization. All methyl-marks seem to be enriched at euchromatic regions of the genome during interphase as DAPI-dense regions being representative for constitutive heterochromatin (pericentric heterochromatin) are not stained. Furthermore nucleoli are excluded. Setd6 co-localizes when comparing the staining patterns. During mitosis the enzyme does not reside on chromatin but is dispersedly distributed around it (fig.3-13). The antibodies for the mono-methyl-marks stain the densely compacted mitotic chromatin (fig. 3-13 and 3-14). The staining for H1K26 mono-methylation is particular as it is not evenly distributed over the chromatin fiber but seems to be concentrated on its surroundings (fig. 3-14). The result that Setd6 and the methyl-mark specific antibodies stain euchromatic (and thereby gene-rich) rather than heterochromatic regions could imply that it is activity is needed for the regulation of genes.

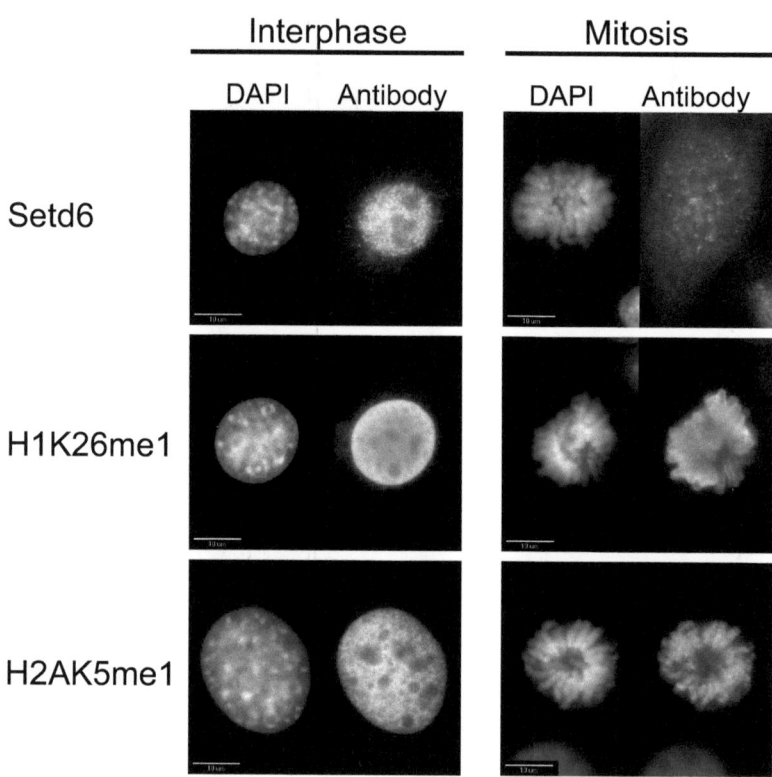

Figure 3-13 During interphase (left panel) Setd6, H1K26me1 and H2AK5me1 stain euchromatic regions of the nucleus. In mitosis (right panel) Setd6 is released from chromatin while the methyl-marks are not removed from histones. Antibodies used: Setd6 #1452; H1K26me1 #270; H2AK5me1 #1919

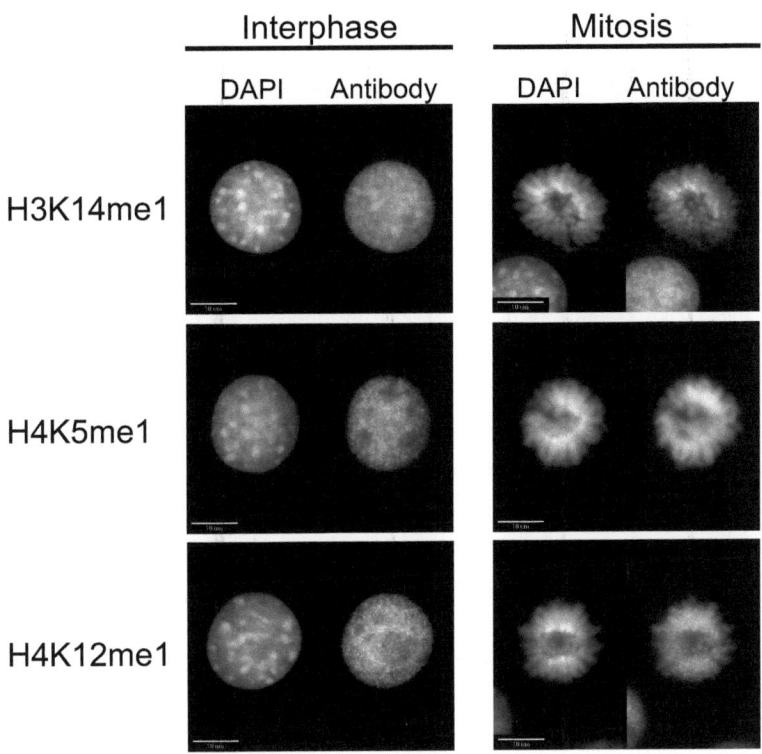

Figure 3-14 During interphase (left panel) H3K14me1, H4K5me1 and H4K12me1 stain euchromatic regions of the nucleus. In mitosis (right panel) the methyl-marks remain on histones. Antibodies used: H3K14me1 #1542; H4K5me1 #1910; H4K12me1 #1543

3.3.9. Setd6 is involved in transcriptional control

To investigate the cellular functions of Setd6 and the associated modifications we established clonal cell lines stably transduced with shRNAs (3 and 13) and confirmed the reduction of Setd6 by real time PCR (fig. 3-15a). For each construct we analyzed several independent cell lines. The cell lines (MR-shRNA 3/9 and MR-shRNA 13/11) with the most significant decrease in the Setd6 level were used for a transcriptional analysis by a custom made in-house cDNA microarray. This microarray was designed using the cDNAs provided in the FANTOM (functional annotation of the mouse) 3 database (http://fantom.gsc.riken.jp/) covering approximately 93% of all protein coding transcriptional units in the murine genome (Carninci et al. 2005; Hayashizaki and Carninci 2006; Maeda et al. 2006). For the analysis we extracted total RNA of wildtype NIH 3T3, Setd6 shRNA 3 clone 9 (MR-shRNA 3/9) and Setd6 shRNA 13 clone 11 (MR-shRNA 13/11) in duplicates. After a quality control of the RNA it was amplified using a T7 polymerase based kit and in the next step labeled with two Alexa dyes (555 or 647) after cDNA generation. The samples were hybridized to the arrays using a hybridization station. The slides were analyzed using a slide scanner and a custom pipeline for quality control. The two biological replicas per cell line were used for 2 technical replicas each. In total for each original cell line 4 microarrays were used. The values of the two technical replicas were averaged. A differential expression in wildtype versus knockdown cells was only taken into account if the p-value (probability for the occurrence of the observed difference by chance) was below 0.05 and if the misregulation was found in both biological replicas. From the resulting list, only genes were analyzed showing at least an up- or downregulation of 2-fold compared to the wildtype cells. In the FANTOM3 library several genes are represented by more than one cDNA. In order to increase the stringency of our analysis we used this redundancy and scored genes only as misregulated if all probes showed the same trend (up- or downregulation).

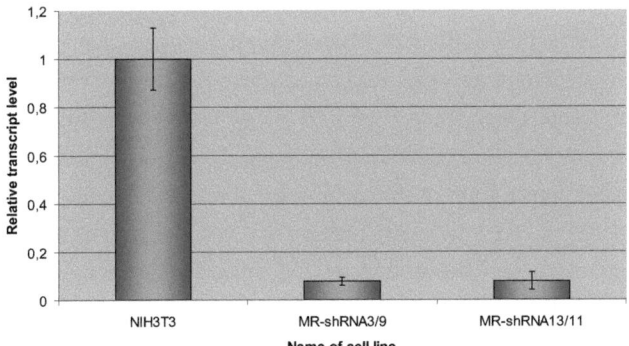

Up regulated genes

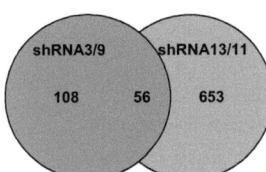

Top 10 genes	Name or domain structure	x fold up
3830417A13Rik	MAGE domain containing uncharac. prot.	20
Tgtp	T-cell specific gtpase	16
Ddah1	Dimethylarginine dimethylaminohydrolase1	11
Gbp4	Macrophage activation 2	9
Ihpk1	Inositol hexaphosphate kinase 1	6
Gpr87	G-protein coupled receptor 87	6
2810409K11Rik	KRAB and znf containing uncharac. prot.	5
0610006I08Rik	Uncharac. transmembrane protein	5
Dach2	Dachshund 2	5
Palld	Palladin, cytoskeletal assoc. protein	5

Down regulated genes

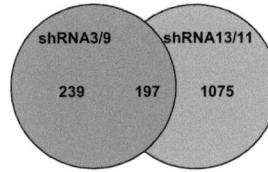

Top 10 genes	Name or domain structure	x fold down
Rspo2	R-spondin 2 homolog	29
Nid2	Nidogen 2	29
Myb	Myeloblastosis oncogene	26
Hmgcs1	3-hydroxy-3-methylglutaryl-coenzyme A s1	20
Hspa4	Heat schock protein 4	20
Lum	Lumican	19
Dda1	DET1- and DDB1-associated protein 1	19
Col3a1	Procollagen, type III, alpha 1	19
Ifi202b	Interferon activated gene 202	18
Gtpbp4	GTP binding protein 4	17

Figure 3-15 a) Stable cell lines with a 90% reduction of the Setd6 mRNA were used for a microarray analysis. b and c) In two independent cell lines 56 genes were consistently upregulated while 197 were downregulated. The tables on the right list the 10 most up- (upper table) or down- (lower table) regulated genes

As a last step we compared the lists of the two independent knockdown cell lines (MR-shRNA3/9 and MR-shRNA13/11) and only if a hit was identified in both we concluded that the change in expression can be correlated with the decrease of the level of Setd6. Using these criteria 56 genes were consistently upregulated and 197 downregulated in the two cell lines (fig. 3-15 b and c). In the list of the downregulated genes 4 are of special interest due to their described role in the literature. Dda1 is an interaction partner of DNA damage-binding protein 1 (DDB1) and might be responsible for the ubiquitination and degradation of histones at the site of DNA lesions induced by UV light (Li et al. 2006). In addition the DDB1 complex is involved in the control of proliferating cells in the developing brain and lens (Cang et al. 2006; Cang et al. 2007). Myb is a transcriptional co-activator and a proto-oncogene being essential for the regulation of haematopoiesis (Greig et al. 2008). Its protein level and activity is regulated by Wnt signalling via Wnt1 (Kanei-Ishii et al. 2004). Hspa4 is important for the cellular response to starvation. Its activity is needed for autophagy by which cells can survive periods of low metabolite influx (Wang et al. 2001). Rspo2 is part of the R-spondin family of proteins which has four members in vertebrates. Those proteins can function in addition to the canonical Wnt molecules as activators for the Wnt signalling pathway and have been implicated in the regulation of developmental processes (Wei et al. 2007). Rspo2 in particular has been reported to be important for patterning in mouse limb development (Aoki et al. 2008). The results indicate that Setd6 has a role in transcriptional regulation but do not allow attributing a repressing or activating function to the binding of Setd6 or its enzymatic activity which would require a analysis by chromatin immunoprecipitation at the promoters of the misregulated genes.

3.3.10. Setd6 interacts with H1 and components of the Wnt pathway

To understand a possible regulation of Setd6 we identified interaction partners of the endogenous protein by immunoprecipitation (IP) using our rabbit polyclonal antibody. With this strategy we were able to circumvent the possibility of non-specific interactions due to an overexpression of a tagged protein. After optimization of the conditions for extraction, binding and washing in immortalized mouse embryonic fibroblasts (iMEFs), we scaled up the number of cells. For the final experiment we used a cell pellet of 3 grams which approximately equals 0.5×10^9 cells (iMEFs). In the first step we covalently coupled the antibody to protein A coated magnetic beads. Cells were harvested by scraping. We prepared cytoplasmic and nuclear (soluble fraction) extracts. Beads and extracts were incubated rotating overnight at 4°C. As control we used beads that were not coupled to an antibody. The next day we collected the beads and washed them with the lysis buffer that was used for the extraction. The bound proteins were eluted by lowering the pH to 2.3 and run on a 4-12% Bis-Tris gradient gel (SDS-PAGE). After silver-staining we excised those bands for mass spectrometry only visible in the actual IP and not in the control. In the cytoplasmic fraction the only detectable band corresponds to the molecular weight of Setd6 which was verified by the mass spectrometric analysis. In the nuclear fraction several bands were unique to the IP. We identified histone H1, Hsp70, Arp2/3, Tip49a and α- and β-catenin as specific interaction partners (fig. 3-16 a and b). Tip49a is a DNA-helicase named after its association with the Tip60 remodelling complex that has been reported to modulate Wnt-signalling by binding to the nuclear β-catenin-TCF complex (T-cell factor) (Feng et al. 2003). α-catenin interacts with β-catenin bound to the intracellular domain of cadherin.

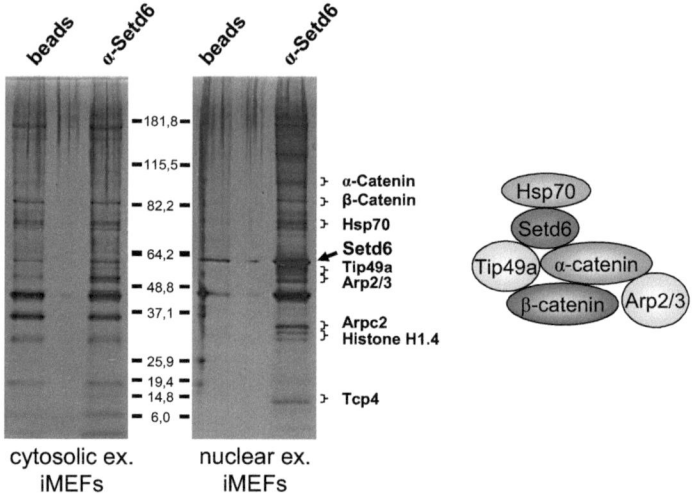

Figure 3-16 a) A Setd6 immunoprecipitation from the nuclear fraction of murine immortalized fibroblasts reveals α- and β-Catenin, Hsp70, Tip49a, Arp2/3, histone H1.4 and Tcp4 as specific interactors. b) Those components could form a complex with Setd6 or interact with the enzyme independently.

This interaction is weakened when due to cell-cell interactions the cadherin organisation is altered which leads to free α-catenin molecules in the cytoplasm inhibiting cell migration, which can be explained by an interaction with Arp2/3 modulating its function as actin assembly factor (Pokutta et al. 2008). α-catenin was also shown to have an antagonistic function in Wnt-signalling to β-catenin, but the mechanism is only partially understood most likely involving a shuttling of α-catenin to the nucleus (Benjamin and Nelson 2008). Hsp70 has a general role as chaperone in concert with other chaperones to stabilize proteins and prevent their aggregation. The gene is induced by heat shock. Histone H1 is the linker histone which is important for chromatin compaction and structure. In the murine genome 7 different isoforms exist which when lost can at least in part compensate for each others function. Only a knockout of three of those genes leads to a phenotypic change in mutant cells shown by transcriptional alterations especially of imprinted genes and a genome-wide change in nucleosome spacing (Fan et al. 2003). Concluding from the mass spectrometry results Setd6 seems to preferentially interact with H1.3 and H1.5.

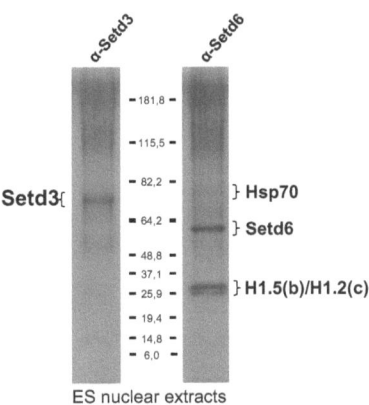

Figure 3-17 An immunoprecipitation of Setd6 from the nuclear extracts of murine embryonic stem cells identifies H1.5 and H1.2 as specific stoichometric interactors.

Having the result for interaction partners in iMEFs we asked whether those interactions are cell line specific or more ubiquitous by immunoprecipitating Setd6 from murine embryonic stem cells using the same protocol as before. As control for this experiment we used an antibody against Setd3 (serum #1457). This protein is highly similar in domain architecture and overall charge being a good and stringent control for specific interactions. In the cytoplasmic fraction we were not able to identify Setd6 specific interactions (data not shown). Interestingly in the nuclear fraction the only persistent interaction in comparison with the IP from iMEFs is between Setd6 and histone H1. The intensity of the silver stained bands suggests that we precipitated both proteins in a one to one stochiometry (fig. 3-17). Methyltransferases for histone H1 have been identified before but a co-immunoprecipitation has not been reported yet and could be an explanation for a targeting mechanism of Setd6 to defined regions of the genome.

3.4. Discussion

Setd6 has broad substrate specificity

In this work, we identify Setd6 as a novel histone lysine mono-methyltransferase by *in vitro* and by *in vivo* experiments. We demonstrate that the enzyme is transcribed throughout murine development and in most adult tissues with high levels in the murine liver, prostate, kidney, salivary gland, heart and thyroid. Setd6 is capable of *in vitro* methylating free recombinant histones H1, H2A, H3 and H4 in HMTase assays. Only towards histone H2B very little activity was found. Such broad substrate specificity has so far not been reported for any SET-domain containing enzyme. This could indicate promiscuous methylation of many or all lysine residues due to recombinant expression of Setd6. Although, histone acetyltransferases such as Tip60 often display similar broad substrate specificity (Kimura and Horikoshi 1998).

Setd6 is a histone lysine mono-methyltransferase

To clarify the issue of the substrate specificity of Setd6 we used a mass spectrometric analysis of N-terminal histone peptides after an HMTase assay. In the spectra only mono-methylated products can be found arguing against promiscuous methylation and for defined substrate recognition. Only for the N-terminus of histone H4 a signal for a doubly-methylated peptide was found in the spectrum. After a MS/MS analysis of all methylated species we were able to identify the specific sites of methylation and found that for H4 two lysines are mono-methylated explaining the before mentioned signal. Setd6 modifies H1 lysine 26 (H1K26), H2A lysine 5 (H2AK5), H3 lysine 14 (H3K14), H4 lysine 5 (H4K5) and H4 lysine 12 (H4K12) by the addition of a single methyl group. As in this assay we offered only the N-termini and could thereby not exclude other sites of methylation in the histone cores we used full length recombinant histones containing point mutations (lysine to alanine substitutions) for the identified residues in an HMTase assay. This enabled us to convincingly demonstrate that the N-terminal lysines are the only ones methylated by Setd6. Mono-methylation at the identified sites has been found in previous analyses (Kubicek S., Jenuwein T. unpublished data). However, the responsible enzyme was unknown. Furthermore, Setd6 is a mono-methyltransferase and prominent examples like H3 lysine 9 and H4 lysine 20 have been demonstrated to be initially mono-methylated and only then can be di- or tri-methylated by other enzymes to reveal their specific function for gene or repeat silencing (Schotta et al. 2004; Karachentsev et al. 2005). While a loss of those enzymes is usually tolerated by the organism in mutant mice studies (eg. Suv39h1 or Suv420h1) a loss of the mono-methylating enzyme as published for Pr-set7 is not viable (Huen et al. 2008). To this point the analysis was based on *in vitro* results. To correlate those results with the *in vivo* enzymatic activity of Setd6 we used a shRNA mediated knockdown of its mRNA an asked by western blot whether the reduction of the enzyme level would correlate with a decrease in the mono-methylation of the identified residues. This correlation was found for H2AK5me1, H4K5me1 and H4K12me1. For H3K14me1 and H1K26me1 the result was not as clear and different time points after transduction or methods for the detection such as mass spectrometry would be advisable. The result is important and allows to conclude that Setd6 is mono-methylating the *in vitro* identified lysines residues also *in vivo* making it the first enzyme with such a broad substrate specificity.

Setd6 localizes to the euchromatic regions in the nucleus

In an immunofluorescence analysis we found that Setd6 localizes to euchromatic regions in the nucleus of murine NIH 3T3 cells. This is in agreement with the notion that we identified a novel histone lysine mono-methyltransferase. In the same experiment we also used antibodies against mono-methylation at the identified residues. From this we can conclude that during interphase their staining is very similar to the one found for Setd6, reinforcing the connection of the enzyme and the identified methyl-marks *in vivo*. This is not the case in mitosis as the marks remain on chromatin while the protein is released and dispersed in the cell. The localization of mono-methylation on histone H1 lysine 26 is especially interesting as it stains the surroundings of the chromatin fibers having the appearance of an envelope. This could be analyzed in future studies possibly using electron microscopy for an in-depth structural analysis, as in earlier studies in *Xenopus laevis* oocytes it was shown that the linker histone is essential for proper condensation of the chromosomes during mitosis (Maresca et al. 2005).

The sites that are mono-methylated by Setd6 have been mainly characterized in their acetylated form being involved in transcriptional regulation (H3K14; H2AK5)(Schiltz et al. 1999; Cuddapah et al. 2009), histone deposition (H4K5; H4K12)(Turner 2000) or mitotic progression (H3K14)(Walter et al. 2008). Even more importantly the histone acetyltransferase Tip60 targets the same residues as Setd6 suggesting an antagonistic role for the two enzymes in e.g. gene regulation or response to different stimuli.

Interaction partners of Setd6 and a possible role in signaling

To understand the biological relevance of Setd6 we decided to look for interactions partners of the enzyme by immunoprecipitation of the endogenous protein using the rabbit polyclonal antibody. In a first attempt we purified the protein and putative binders from immortalized murine fibroblasts (iMEFs) and identified histone H1, Hsp70, Arp2/3, Tip49a and α- and β-catenin by mass spectrometry in the nuclear fraction as specific interactors. The interaction with H1 was not surprising as it is a substrate of the enzyme and Hsp70 is a chaperone stabilizing misfolded proteins which often appear during purification. Tip49a is a DNA-helicase and a common subunit in chromatin associated complexes. Importantly it is part of the Tip60 complex acetylating the residues in the histone tails (Squatrito et al. 2006) that are methylated by Setd6 again reinforcing the idea of an antagonistic role. α-catenin has been shown to interact with β-catenin and Arp2/3 (Benjamin and Nelson 2008) in earlier studies but those interactions are mutually exclusive, whether they are in a common complex with Setd6 remains to be clarified in future experiments. Another important aspect is that β-catenin is an important molecule for Wnt signalling as it shuttles to the nucleus and serves as a co-activator for transcription factors in this pathway. α-catenin has an antagonistic role for this cascade. How Setd6 is involved in this regulation could be analyzed in future studies. Tip60 on the other has already been characterized for its role in different signaling pathways such as Wnt (Feng et al. 2003) or Notch (Kim et al. 2007) signaling. In future experiments it should be investigated whether the Setd6 complex and the Tip60 complex are targeting the same genes in the genome and are antagonizing each other in the regulation of those (fig. 3-18).

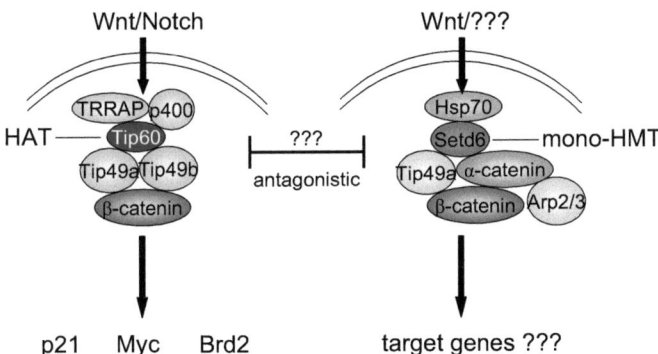

Figure 3-18 The Tip60 complex has been implicated in the regulation of downstream targets of the Wnt and Notch signaling pathway. The interaction of Setd6 with β-Catenin and the downregulation of Rspo2 in Setd6 knockdown cells could hint to an antagonistic role in those pathways as both enzymes target the same residues in the histone tails.

To understand the role of Setd6 and the associated marks in epigenetic regulation we established stable Setd6 knockdown NIH 3T3 cell lines (MR-shRNA 3/9 and MR-shRNA 13/11) using shRNA 3 and shRNA 13 and analyzed their transcriptome using a cDNA based microarray. After a stringent filtering of the data we found that 56 genes are consistently up- and 197 genes are consistently downregulated. Interestingly in the list of the ten most downregulated genes we found R-spondin 2 (Rspo2). In the murine genome four R-spondins are encoded and is has been demonstrated that they can function as an alternative to Wnt ligands for the activation of the Wnt-pathway by the binding to the associated receptors (Lrps6 and Frizzled8).(Nam et al. 2007) In this way Setd6 could modulate the signaling response which should be examined by follow up experiments especially considering the established role of Tip60 in the Wnt-signaling cascade. A second interesting gene from the list of downregulated genes is Hspa4 as it important for the response to low metabolite levels in the cells (Wang et al. 2001). Setd6 could be regulating this response on a transcriptional level. The analysis did not allow concluding whether Setd6 and the associated methyl-marks have a repressive or activating function. This should be addressed by directed chromatin immunprecipitations at the identified misregulated genes.

Setd6 has a proposed role in the regulation of pluripotency

The Tip60 enzyme has been implicated in the maintenance of pluripotency in murine ES cells, as a loss of components of the Tip60 complex lead to a partial differentiation of the cells (Fazzio et al. 2008). Based on our data demonstrating that Setd6 and Tip60 target the same residues we propose that a loss of Setd6 could have the inverse effect by making the cells less sensitive to differentiation. This could be examined by a knockout of the gene or a shRNA mediated knockdown followed by lineage specific differentiation or simple withdrawal of factors needed for pluripotency in the media (e.g. LIF) (fig. 3-19 a, b and c). It would be also interesting to see whether Setd6 knockout ES cells could go germline and produce viable mice or whether the developmental program would be aborted at a specific stage due to impaired differentiation

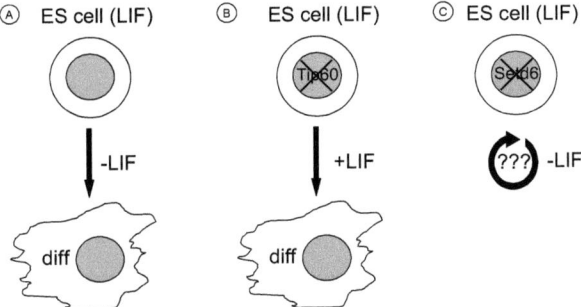

Figure 3-19 a) The removal of LIF from the media of embryonic stem cells leads to their differentiation. b) A knockdown of components of the Tip60 complex has a similar effect. c) A Setd6 knockdown could have the opposite effect keeping the cells in a pluripotent state even after the removal of LIF

Importantly in the list of the 10 most downregulated genes in the Setd6 knockdown cell lines we identified Myb. This protein has an established role in haematopoiesis where it regulates proliferation and has an inhibitory effect on differentiation, its activity is regulated via the Wnt signaling pathway (Greig et al. 2008). A second gene from this list is Dda1. This protein is an interacting partner of DNA damage-binding protein 1 (DDB1) and thereby involved in the response to DNA damage after e.g. UV treatment (Li et al. 2006). More importantly the DDB1 complex has been shown to be important for the regulation of proliferation of cells in the developing brain and lens (Cang et al. 2006). The Setd6 dependent regulation of those genes in NIH 3T3 cells strengthens the idea of an involvement in differentiation or pluripotency.

We can conclude that we identified a novel nuclear histone lysine monomethyltransferase involved in epigenetic regulation methylating residues that had only been characterized in their acetylated form. The enzyme has a broad tissue distribution but seems to be rather involved in the regulation of a defined set of genes and not in general gene activation or inactivation. Its interaction with components of the Wnt signalling pathway and the misregulation of Rspo2 as a modifier of this pathway in the knockdown cell lines might indicate a role for

Setd6 in signalling. Furthermore the strong interaction with histone H1 could explain the targeting mechanism of Setd6 and its nuclear localization as in its sequence no nuclear localization signal (NLS) can be found. This notion is strengthened by an immunofluorescence analysis of Setd6 in HeLa cells revealing that the enzyme can be cytoplasmic or nuclear depending on the cell density (data not shown). Histone H1 has also been shown to be capable of shuttling from the nucleus to the cytoplasm when cells undergo apoptosis (Yan and Shi 2003). We can also conclude that an antagonistic role with the histone acetyltransferase Tip60 is very likely as they target the same residues and both seem to function in the same signaling pathway. To clarify many of the open questions we currently prepare a conditional gene knockout and perform in depth ChIP analyses.

3.5. Additional results for Setd6

3.5.1. An increased level of Setd6 leads to the formation of micro nuclei

To dissect the protein function of Setd6 from its enzymatic activity, we used forced overexpression of wildtype and mutant Setd6. We subcloned the cDNA of Setd6 into a lentiviral vector (pLenti4TNT1Puro/Setd6). A transduction with viral particles produced from this plasmid leads to an overexpression of the protein containing an N-terminal Flag-HA (haemaglutinin) tag and a C-terminal enhanced green fluorescent protein (eGFP) tag. We made such construct for the wildtype Setd6 protein and the mutant protein (pLenti4TNT1Puro/Setd6mut; Y285A). 48 hours after transduction of NIH 3T3 cells with the virus, we sorted single GFP-positive cells into 96-well tissue culture plates to derive clonal cell lines. We analyzed those cells for their fluorescent signal and used three cell lines overexpressing the wildtype protein and three overexpressing the mutant at equal levels for an immunofluorescence analysis. For this analysis we stained the cells with DAPI and two antibodies, one recognizing Setd6 and the other the eGFP tag. This indicates that the cell lines with an increased level of the enzyme formed micronuclei in a high percentage (>10%) of the cells. By contrast, wildtype cells or cells overexpressing the mutant protein displayed micronuclei in a low number of cells (approx. 2%) (fig. 3-20 a and b). Micronuclei are formed when a chromosome or chromosome fragment is not properly segregated during mitosis, thereby allowing an estimate of overall chromosomal stability. Although a difference between wildtype and mutant can be found, the degree to which the induction of micronuclei can be directly linked to the protein-function or the enzymatic activity towards histones remains to be examined in more detail.

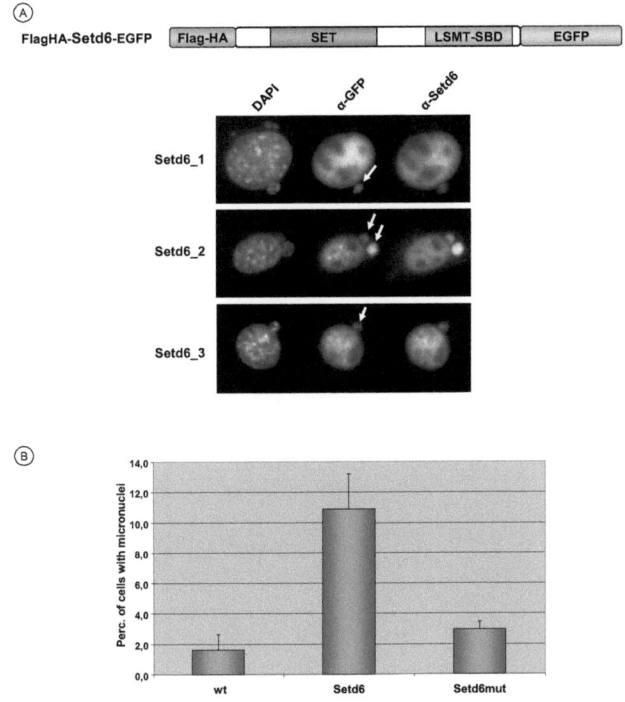

Figure 3-20 Stable Setd6 overexpressing cells have a high percentage of micro nuclei (a) when compared with wildtype or Setd6 mutant overexpressing cells (NIH3T3) (b).

An impact on gene regulation could not be found by analyzing total RNA samples of those cell lines in the same way as described for the knockdown cell lines (see page 40) indicating that the induction of micronuclei is most likely influenced by interactions with other proteins and not by a change in the transcriptional level of certain genes.

3.5.2. Setd6 methylates non-histone proteins

We could demonstrate that the methylation of histones by Setd6 is site-specific and also occurs in cells. As Setd6 has such broad substrate specificity towards histones, we asked whether non-histone substrates could be identified by using the amino acid sequence of the known sites for an alignment with the mouse proteome. For this purpose we used the BLAST algorithm (http://www.expasy.org/tools/blast/) increasing the E threshold (number of expected hits in a random database) to 10000, not filtering for low-complexity regions and using the amino acid sequence of histones H3 and H4 positioning the methylated lysine in the middle and adding the 5 amino acids N-terminal and C-terminal.

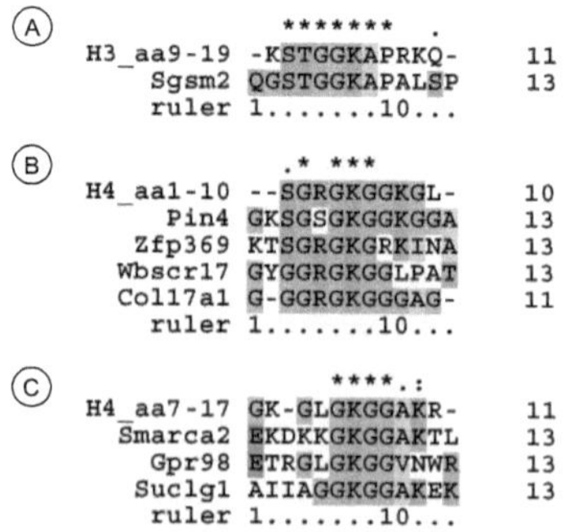

```
(A)              *******  .
      H3_aa9-19 -KSTGGKAPRKQ-    11
          Sgsm2 QGSTGGKAPALSP    13
          ruler 1.......10...

(B)               .* ***
      H4_aa1-10 --SGRGKGGKGL-    10
           Pin4 GKSGSGKGGKGGA    13
         Zfp369 KTSGRGKGRKINA    13
        Wbscr17 GYGGRGKGGLPAT    13
        Col17a1 G-GGRGKGGGAG-    11
          ruler 1.......10...

(C)              ****.:
      H4_aa7-17 GK-GLGKGGAKR-    11
        Smarca2 EKDKKGKGGAKTL    13
          Gpr98 ETRGLGKGGVNWR    13
         Suclg1 AIIAGGKGGAKEK    13
          ruler 1.......10...
```

Figure 3-21 Following a BLAST search using the sequences of a) histone H3 aa9-19 b) histone H4 aa1-10 and c) H4 aa7-17 we retrieved sequences from non-histone proteins that could be potential substrates for Setd6.

We expected to retrieve peptides sequences that would have a lysine residue and would have an overall high similarity to the peptide initially used for the BLAST search. Based on this analysis we identified 8 potential peptide stretches in non-histone substrates that could be targets of Setd6. Candidate proteins with high-similarity in peptide stretches were Sgsm2 (aa112-124), Pin4 (aa5-17), Wbscr17 (aa89-102), Col17a1 (aa433-443), Zfp369 (aa619-631), Gpr98 (aa5092-5104), Suclg1 (aa289-301) and Smarca2 (aa998-1010) (fig. 3-21 a, b and c). In a radioactive HMTase assay we used synthetic peptides including an H4 peptide (aa 1-23) as positive control to determine whether recombinant Setd6 would have catalytic activity towards the lysines. As expected a pronounced signal was detectable for H4 and interestingly a signal of similar intensity was found for Wbscr17 and Zfp369. Smarca2, Gpr98 and Pin4 were methylated to a minor degree. No radioactivity was incorporated for Suclg1, Col17a1 and Sgsm2 (fig. 3-22).

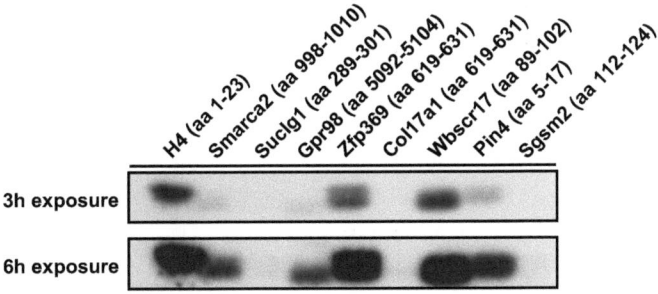

Figure 3-22 Peptides of non-histones proteins with a high similarity to the histones targets of Setd6 have been used in an HMTase assay revealing Zfp369K625 and Wbscr17K95 as substrates for the enzyme.

Zfp369 is a transcription factor containing two KRAB (krueppel associated box) domains separated by one SCAN (leucine rich region) domain and 2 C-terminal zinc-fingers. It has been described that the protein interacts with the neurotrophin receptor p75^{NTR} and has a high similarity with Nrif1 (neurotrophin receptor interacting factor) even being able to partially compensate for a loss of this protein in a mouse knockout model. An overexpression of the protein leads to a cell cycle arrest (Benzel et al. 2001). The methylated lysine is in the region between the second KRAB domain and the first zinc-finger (fig. 3-23). Wbscr17 (Williams-Beuren syndrome chromosomal region 17 protein homolog) is a putative polypeptide N-acetylgalactosaminyltransferase catalyzing the transfer of an N-acetyl-D-galactosamine group to a serine or a threonine residue which is the initial reaction in the O-linked oligosaccharide biosynthesis. The human homolog of this gene is deleted in the so called Williams syndrome, a developmental disorder. The protein contains a glycosyl transferase (family 2) domain responsible for the catalytic activity and a ricin B lectin domain capable of binding sugar moieties.(Merla et al. 2002; Nakamura et al. 2005) The modified lysine is located in the N-terminal part (fig. 3-23).

Smarca2 (SWI/SNF related, matrix associated, actin dependent regulator of chromatin, subfamily a, member 2) is an ATP-dependent DNA helicase and a component of the SWI/SNF remodelling complex functioning as a transcriptional coactivator. It has a C-terminal DNA-helicase domain flanked by a bromodomain and an N-terminal QLQ (glutamine-leucine-glutamine) protein interaction domain. Its proper function is important for cells to be able to enter the quiescent state by regulating the level of cyclin A (Coisy-Quivy et al. 2006). A decrease in the protein level is frequent in gastric cancer types (Yamamichi et al. 2007). In complex with CCAAT/enhancer-binding protein alpha (C/EBPalpha) it is responsible for a reduction in the proliferative capacity of the liver in old animals by the inhibition of E2F-dependent promoters (Itoh et al. 2008). Setd6 potentially methylates a lysine in a conserved region amongst SNF2 family members in the middle of the protein (fig. 3-23).

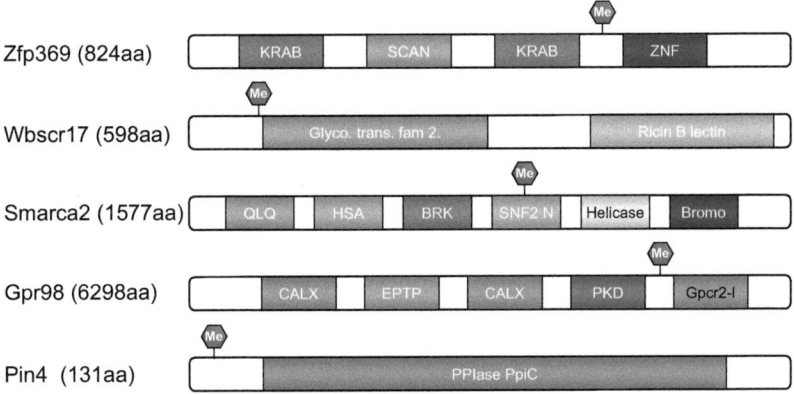

Figure 3-23 From the HMTase assay Setd6 can methylate 5 of the 8 tested peptides from non-histone substrates. Depicted are the domain architecture of those proteins and the putative sites of methylation.

Gpr98 (G protein-coupled receptor 98) has been demonstrated to be the gene responsible for seizures in response to tonic stress in the Frings epilepsy mouse model which can be linked to malformation of stereocilia in the mutant mice. The protein is integral to the plasma membrane and has several calx-beta motifs found in sodium calcium exchangers (Skradski et al. 2001). The methylation targets a lysine in the C-terminal part of Gpr98 (fig. 3-23). Pin4 ((peptidyl-prolyl cis/trans isomerase) NIMA-interacting, 4) is a peptidyl-prolyl cis/transisomerase accelerating protein folding and is probably methylated in its very N-terminus (fig. 3-23). The result of the analysis indicates that Setd6 might have several non-histone targets and could regulate processes in the cell without directly affecting histone/chromatin biology. Whether the enzymatic activity can also been found towards the full-length proteins and whether and when it occurs *in vivo* are questions that have to be addressed in further experiments.

3.6. Discussion of additional results

We examined how a forced overexpression of wildtype or the mutant Setd6 protein would affect NIH 3T3 cells. We asked for obvious phenotypic changes and found that the overexpression of the wildtype protein correlates with a high frequency of micronuclei in the cells being a general marker for damaged chromosomes and genome instability often caused by genotoxic compounds or radiation leading to missegregated chromosomes during mitosis. Although this finding is dependent on the enzymatic activity of Setd6 protein function versus enzymatic activity needs to be explored further. How the increased level of the protein causes the defect has to be investigated in detail as a transcriptome analysis revealed no major misregulations. A possibility would be the methylation of a non-histone substrate causing an altered response to DNA damage or checkpoint control. This led us to the last experiment were we used the sequence of the amino acid stretches surrounding the methylated lysines in the tails of histone H3 and H4 to search for similar peptides in the murine proteome expecting to identify non-histone substrates of Setd6. By this approach we identified 8 putative targets of which two turned out to be substrates of equal quality compared to the N-terminus of histone H4 in an HMTase assay. The peptides are fragments of zinc finger protein 369 (Zfp369) and Williams-Beuren syndrome chromosomal region 17 protein (Wbscr17). This result is a first indication that Setd6 has non-histone substrates and could thereby regulate cellular processes in addition to directly binding to chromatin and methylating histones interfering with transcriptional regulation. In future analyses it should be examined whether Setd6 is capable of methylating the full length proteins and how an overexpression or knockdown of the enzyme affects for example their localization or enzymatic activity in cells.

4. Lysine methylation on p53

4.1. Introduction

4.1.1. Post-translational modifications of p53

p53 is a transcription factor regulating the transcription of many genes important for a response to stresses such as DNA damage, misexpressed oncogenes or low metabolites thereby inducing apoptosis, cell cycle arrest or senescence (Bode and Dong 2004; Toledo and Wahl 2006). Its functions are to prevent the cell from progressing through the cell cycle when damage occurs disturbing the integrity of the genome and to decide whether this damage is too severe (apoptosis) or can be repaired (cell cycle arrest). TP53, the gene encoding p53, is mutated in approximately 50% of all human cancers usually rendering the protein inactive (Pfeifer and Holmquist 1997). In an additional 10-20% of the other half MDM2 or MDM4, being inhibitors of the transactivation function of p53, are overexpressed (Toledo and Wahl 2006). This implies that a tight regulation of this protein is needed as a decreased activity makes an organism prone to tumor development while an increase leads to massive cell death or impaired cell division. Similar to many other proteins involved in the response to defined stimuli its activity is modulated by post-translational modifications affecting its stability, its DNA-binding affinity or its interaction with other proteins. For the human protein more than 20 phosphorylation sites have been described in the literature (Bode and Dong 2004). Some of them when phosphorylated increase the transactivation potential often through interfering with the binding of MDM2, which otherwise ubiquitinates p53 and leads to its degradation. The sites are in part targeted by several kinases and some kinases phosphorylate more than one residue allowing a certain redundancy in the regulation and a well defined output. While the N-terminal transactivation domain (TAD) contains many of the phosphorylated sites the C-terminal domain (CTD) was reported to be modified by ubiquitination, sumoylation, acetylation or methylation at different lysine residues (Bode and Dong 2004; Chuikov et al. 2004; Huang and Berger 2008). The addition of ubiquitin moieties to lysine residues in the CTD is a signal for nuclear export and degradation of p53 and is needed for its steady-state level allowing a well-controlled pool of protein that can be rapidly activated as a stress response (Clegg et al. 2008). In this case p53 is phosphorylated at different residues in the TAD reducing the binding affinity of MDM2 and favouring the interaction with the acetyltransferase p300 acetylating lysine residues in the CTD and so blocking ubiquitination. The acetylation thereby stabilizes p53 and increases specific DNA binding. At target promoters p300 acetylates histones and helps inducing transcription. P300/CBP-associated factor (PCAF) is a second acetyltransferase also modifying and activating p53 (Sterner and Berger 2000). The experimental evidence shows that p53 activity increases when the protein becomes acetylated. In agreement with these results is that deacetylation by SIRT1 reduces the transcriptional activation of p53 target genes and impairs the cellular response to γ-irradiation (Vaziri et al. 2001; Deng 2009). Many proteins localize to the promyelocytic leukaemia (PML) nuclear bodies (NBs) and form a dynamic subnuclear multiprotein complex containing components important for the regulation of transcriptional activity (e.g. acetyltransferases, deacetylases or the PML tumour-suppressor protein) including p53 (Gottifredi and Prives 2001). The localization to this structure requires the addition of a polypeptide by the enzyme small ubiquitin-like modifier 1 (SUMO1). For p53 sumoylation seems to have a repressive function and it is not clarified yet whether the protein can be desumoylated and which enzyme would catalyze this reaction (Bode and Dong 2004; Toledo and Wahl 2006). Histone lysine-methyltransferases (HMTases) are involved in transcriptional regulation by altering the accessibility of the DNA double-helix for components of the transcriptional machinery either directly or via the recruitment of chromatin binding proteins. In the murine genome approximately 50 of such enzymes exist and for several the enzymatic activity has been at least in part described in the literature. While in the beginning a clear focus was on their histone modifying activity, it became obvious that also non-histone targets are important for the final output and to understand their involvement in response to pathway activation. For p53 it has been demonstrated that methylation by the HMTase Set9 at lysine 372 in the CTD positively affects its transactivation function for the p21 gene, increases its stability and is important for induction of apoptosis after DNA-damage (Chuikov et al. 2004; Ivanov et al. 2007). Also Prset7, being one of the few reported histone lysine mono-methyltransferases with an important function in development, has been identified as an enzyme modifying p53. The targeted residue is lysine 382 and in contrast to the Set9 mediated modification this addition of a methyl group leads to a reduced induction of the p21 gene and the p53 up-regulated modulator of apoptosis (PUMA) gene. A knockdown of the enzyme leads to an increase of apoptosis after DNA-damage and increased amounts of DNA-bound p53.(Shi et al. 2007) In this work we present data demonstrating that mono-methylation of p53 at lysine 370 by SET and MYND domain-containing protein 2 (Smyd2) has a repressive function while di-methylation by a still unidentified HMTase at the same site is activating as it is needed for the recruitment of the coactivator p53-binding protein 1 (53BP1). The removal of this methyl-mark is catalyzed by Lysine-specific histone demethylase 1 (LSD1), which we characterize for its repressive function on p53 activity. In addition we provide evidence for a cross-talk between the activating p53K372 methylation by Set9 and the repressive p53K370 methylation.

4.2. Repression of p53 activity by Smyd2-mediated methylation

Jing Huang[1]*, Laura Perez-Burgos[2]*, Brandon J. Placek[1], Roopsha Sengupta[2], Mario Richter[2], Jean A. Dorsey[1], Stefan Kubicek[2], Susanne Opravil[2], Thomas Jenuwein[2] & Shelley L. Berger[1]

Specific sites of lysine methylation on histones correlate with either activation or repression of transcription[1-3]. The tumour suppressor p53 (refs 4-7) is one of only a few non-histone proteins known to be regulated by lysine methylation[8]. Here we report a lysine methyltransferase, Smyd2, that methylates a previously unidentified site, Lys 370, in p53. This methylation site, in contrast to the known site Lys 372, is repressing to p53-mediated transcriptional regulation. Smyd2 helps to maintain low concentrations of promoter-associated p53. We show that reducing Smyd2 concentrations by short interfering RNA enhances p53-mediated apoptosis. We find that Set9-mediated methylation of Lys 372 inhibits Smyd2-mediated methylation of Lys 370, providing regulatory cross-talk between post-translational modifications. In addition, we show that the inhibitory effect of Lys 372 methylation on Lys 370 methylation is caused, in part, by blocking the interaction between p53 and Smyd2. Thus, similar to histones, p53 is subject to both activating and repressing lysine methylation. Our results also predict that Smyd2 may function as a putative oncogene by methylating p53 and repressing its tumour suppressive function.

Members of the Smyd (SET and MYND domain) family and Suv4-20 were expressed in bacteria and tested for histone and p53 methylation. Smyd2, but not Smyd3, Smyd5 or Suv4-20, methylated a peptide spanning amino acid residues 358–393 in the carboxy terminus of p53 (Supplementary Fig. S1a). Peptide mapping showed that lysine 370 (K370) is the methylation site for Smyd2 (Fig. 1a, b). Unmodified, mono-, di- and tri-methylated K370 p53 peptides (residues 361–380) were subjected to Smyd2 methylation in vitro followed by mass spectrometry (Supplementary Fig. S1b). The results showed that Smyd2 mono-methylates K370 in p53 in vitro.

Polyclonal antibodies specifically recognizing mono-methylated (p53K370me1), di-methylated (p53K370me2) and tri-methylated (p53K370me3) K370 were generated and tested for specificity in vitro

(Supplementary Fig. S2a). These antibodies were used to test p53 methylation by Smyd2 in vivo. H1299 cells were transfected with Flag–p53, Flag–p53(K370A) or Flag–p53(K370R) (Fig. 2a and Supplementary Fig. S2b), either with or without Flag–Smyd2 or a methylation-defective mutant Flag–Smyd2(MD) (Supplementary Fig. S3). Flag immunoprecipitates were subjected to western blot analysis with methylation-specific antibodies. Flag–p53, but not the K370R mutant, was detected by the antibody to p53K370me1 (Fig. 2a, compare lanes 1 and 4). The extent of K370 methylation of Flag–p53, but not the K370R mutant, was greatly increased by cotransfection with Flag–Smyd2 (Fig. 2a, compare lanes 2 and 5). Detection of p53 by the antibody to p53K370me1 was greatly diminished by

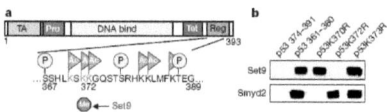

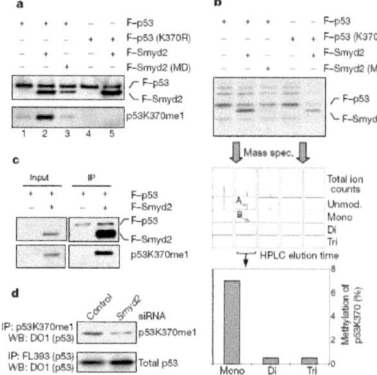

Figure 1 | Smyd2 methylates p53 at K370 in vitro. a, p53 functional domains. TA, transactivation; Pro, proline-rich; Tet, tetramerization; Reg, regulatory; Ac, acetylation; P, phosphorylation; Me, methylation. Set9 methylates p53 at K372. **b**, Autoradiogram of HMTase assay with recombinant Set9 (top) or Smyd2 (bottom) on p53 peptide 374–391, 361–380 or 361–380 bearing K370R, K372R or K373R substitutions.

Figure 2 | Smyd2 methylates p53 at K370 in vivo. a, Western blot analysis of Flag immunoprecipitates from H1299 cells expressing Flag–p53 (lanes 1–3) or Flag–p53(K370R) (lanes 4–5) either alone (lanes 1 and 4) or with Flag–Smyd2 (lanes 2 and 5) or Flag–Smyd2(MD) (lane 3). **b**, Coomassie staining (top) and mass spectrometry analysis (middle and bottom) of Flag immunoprecipitates from H1299 cells. **c**, Western blot analysis of input and Flag immunoprecipitates from U2OS cells. **d**, Western blot analysis of U2OS cells transfected with control or Smyd2 siRNA, followed by immunoprecipitation with antibody to p53K370me1 (top) or to full-length p53 (FL393; bottom).

[1]Gene Expression and Regulation Program, The Wistar Institute, Philadelphia, Pennsylvania, 19104, USA. [2]Research Institute of Molecular Pathology (IMP), The Vienna Biocenter, Vienna, A-1030, Austria.
*These authors contributed equally to this work.

cotransfection with Flag–Smyd2 (MD) (Fig. 2a, compare lanes 2 and 3). The antibody to p53K370me2 recognized p53 largely without regard to its methylation status at K370, whereas the antibody to p53K370me3 did not detect any tri-methylation signal (data not shown).

The methylation status of Flag-p53 in H1299 cells was also tested by mass spectrometry (Fig. 2b). Flag-p53 was immunoprecipitated and gel-isolated after Coomassie blue staining. The results showed that p53 is mono-methylated by Smyd2 in vivo (Fig. 2b).

We analysed methylation of p53 in U2OS cells. Cotransfected Flag-p53 and Flag-Smyd2 yielded mono-methylated p53 (Fig. 2c), but not di- or tri-methylated p53 (data not shown). The ability of endogenous Smyd2 to methylate endogenous p53 was tested. The amount of p53K370me1 was reduced when cells were treated with a short interfering RNA (siRNA) targeting Smyd2 as compared with a control siRNA (Fig. 2d; see Supplementary Fig. S4b, c, for siRNA controls). These in vivo data support our in vitro findings that Smyd2 mono-methylates p53 at K370.

To address the role of Smyd2 in regulating the function of p53, we used the human fibroblast cell line BJ-DNp53, which stably expresses a dominant-negative variant of p53 to inactivate the function of endogenous p53 (ref. 9). Reducing Smyd2 concentration with Smyd2 siRNA correlated with increased p21 and mdm2 expression in control BJ cells but not in BJ-DNp53 cells (Fig. 3a and Supplementary Fig. S4a) with or without irradiation. Smyd2 reduction also resulted in an increase in mRNA (Supplementary Fig. S4b) and protein (Supplementary Fig. S4c) concentrations of p21 and mdm2 in U2OS cells. However, this effect was muted in H1299 cells, which are p53 null (Supplementary Fig. S4c). Taken together, these results show that Smyd2 negatively regulates p53-responsive genes in a p53-dependent manner.

To determine whether Smyd2 regulates the function of p53 specifically through K370 methylation, wild-type p53, p53(K370R) or an empty expression vector was transfected into H1299 cells stably expressing control or Smyd2 short hairpin RNA (shRNA; Fig. 3b). Wild-type p53 induced p21 expression to a greater extent in cells expressing Smyd2 shRNA than in cells expressing control shRNA. However, p53(K370R) showed the same transcriptional ability in both cell lines, and this ability matched the higher activity of wild-type p53 in H1299 cells expressing Smyd2 shRNA (Fig. 3b). These findings indicate that Smyd2 may downregulate the transcriptional activation ability of p53 through K370 methylation.

We investigated whether the p53K370me1 modification occurs on p53 associated with cognate genes. Double chromatin immunoprecipitation (ChIP) was done with Flag antibody for the first ChIP, followed by Flag peptide elution, and then p53K370me1 antibody for the second ChIP. The total p53 signal decreased when Smyd2 was cotransfected, whereas the p53K370me1 ChIP essentially remained the same (data not shown). Thus, the ratio of p53K370me1 signal to total p53 signal increased in response to cotransfected Smyd2 (Fig. 3c, compare lanes 1 and 2). This increase was eliminated in the Smyd2(MD) mutant (Fig. 3c, compare lanes 2 and 3). The higher ChIP signal required intact K370, because the increase was not seen in Flag-p53(K370R) immunoprecipitates (Fig. 3c, compare lanes 1–3 and lanes 4–6). The higher signal of p53K370me1, as compared with IgG, arising from p53(K370R) immunoprecipitates presumably results from the residual recognition of unmodified p53 by the methylation antibody when p53 is overexpressed (Fig. 3c, lanes 4–6).

The endogenous amounts of total p53 and K370-methylated p53 bound to the p21 promoter were assessed by ChIP assay in U2OS cells (Supplementary Fig. S5a). Although the promoter-associated amount of total p53 increased markedly on adriamycin treatment, K370-methylated p53 increased only slightly (Supplementary Fig. S5a), resulting in a decrease in the relative percentage of K370-methylated p53 associated with the p21 promoter (Supplementary Fig. S5b). These data further support the view that Smyd2-mediated K370 methylation is a repressive modification of p53.

We investigated the mechanism underlying the repressive effect of Smyd2-mediated K370 methylation. We found no change in the distribution of p53 between the nucleus and cytoplasm when Smyd2 was reduced (Supplementary Fig. S6). In addition, we detected no change in the amount of total p53 (Supplementary Figs S4c, S7 and S8d), K382-acetylated p53 and S15-phosphorylated p53 (Supplementary Fig. S7) when Smyd2 was reduced or increased.

Next, we examined the effect of increasing Smyd2 concentrations on p53 binding at the p21 gene by ChIP using U2OS cell lines inducibly expressing either empty vector or Smyd2 (Supplementary Fig. S8a). Increasing Smyd2 lowered the amount of p53 bound to the p21 promoter (Supplementary Fig. S8b). The decreasing amount of promoter-associated p53 correlated with decreasing amounts of p21 mRNA (Supplementary Fig. S8c) and p21 protein (Supplementary

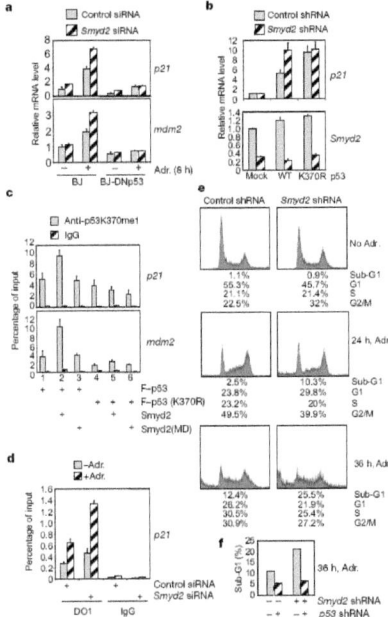

Figure 3 | Smyd2 represses the function of p53. a, Real-time PCR analysis of p21 and mdm2 mRNA in BJ and BJ-DNp53 cells transfected with control or Smyd2 siRNA by mock or 8-h adriamycin treatment. b, Real-time PCR analysis of relative Smyd2 and p21 mRNA in H1299 cells expressing control shRNA or Smyd2 shRNA and transfected with an empty, p53 or p53(K370R) expression vector. c, Sequential ChIP assay (Flag ChIP followed by p53K370me1 ChIP) to assess recruitment of p53K370me1 to the promoters of p53-responsive genes in H1299 cells. d, ChIP assay of U2OS cells transfected with control or Smyd2 siRNA followed by mock or adriamycin treatment. e, Flow cytometry analysis of U2OS cells stably expressing control or Smyd2 shRNA, followed by adriamycin treatment. f, U2OS cells stably expressing control or Smyd2 shRNA were transduced with lentivirus bearing luciferase control or p53 shRNA. Data are the mean ± s.d. (a–d).

Fig. S8d). We also reduced Smyd2 by siRNA and observed an increase in promoter-associated p53 (Fig. 3d). These data suggest that K370-methylation of p53 reduces the DNA-binding efficiency of p53.

To address the role of Smyd2 in p53-mediated cell-cycle arrest and apoptosis, we carried out flow cytometry analysis in U2OS cells stably expressing either control or *Smyd2* shRNA with or without adriamycin treatment (Fig. 3e). Reduction of Smyd2 did not cause obvious changes in the apoptotic fraction of cells in U2OS cells under the non-DNA damage condition (Fig. 3e, top). However, 24 h after DNA damage, fivefold more cells expressing *Smyd2* shRNA (10.3%) entered apoptosis (sub-G1) than did cells expressing control shRNA (2.5%; Fig. 3e, middle). This difference remained after 36 h of adriamycin treatment (Fig. 3e, bottom).

To address whether the effect of Smyd2 on apoptosis is p53 dependent, we reduced the amounts of p53 protein and *p53* mRNA in U2OS cells stably expressing control or *Smyd2* shRNA (Supplementary Figs S9 and S4c). Ablation of p53 almost completely eliminated the difference in apoptosis between U2OS cells expressing control shRNA and those expressing *Smyd2* shRNA (Fig. 3f). Because reintroduction of p53 into H1299 is known to induce apoptosis[10], we transfected an empty or p53 expression vector into H1299 cells expressing control or *Smyd2* shRNA cells and subjected them to γ-irradiation (Supplementary Fig. S10). We found ectopically expressed p53 caused modest cell apoptosis without DNA damage (Supplementary Fig. S10, top, compare lanes 3 and 4 to lanes 1 and 2). On DNA damage, the addition of p53 resulted in more apoptosis in H1299 cells expressing *Smyd2* shRNA than in those expressing control shRNA (Supplementary Fig. S10, bottom, compare lanes 4 and 3). We were unable to use this p53 overexpression approach in H1299 cells to test further the prediction that the K370R mutant would cause even more cell death, because the apoptosis caused by overexpressed p53 was already severe. Taken together, we conclude that Smyd2 is involved in p53-dependent cell-cycle arrest and apoptosis.

We examined the possible interplay between the activating modification p53K372me1 (ref. 8) and the repressing modification p53K370me1. We first tested whether there is any mutual effect on enzyme activities by using peptides and recombinant enzyme activities. Methylation by Set9 was not altered on peptides that were mono- or di-methylated at K370 (Fig. 4a, top). We also tested the ability of Set9 to mono-, di- or tri-methylate p53 peptides at K372, an issue that has not been reported[8]. We found that Set9 can mono- and di-methylate p53 at K372 *in vitro* (Supplementary Fig. S10a), and that methylation by Smyd2 is lowered on both K372me1 and K372me2 peptides as compared with unmodified peptides (Fig. 4a, bottom). Taken together, our *in vitro* data suggest a 'one-way' cross-talk, in which K372 methylation inhibits K370 methylation.

We therefore tested whether there is a mutual effect *in vivo*. Whereas higher concentrations of p53K372me1 resulted in greatly reduced amounts of p53K370me1, higher concentrations of p53K370me1 did not lower the basal amounts of p53K372me1 (Fig. 4b, top). Furthermore, even the great increase in p53K372me1 obtained by cotransfection with Set9 was not lowered by Smyd2 cotransfection (Fig. 4b, bottom).

A trivial explanation for the observed Set9/K372me1 inhibition of p53K370me1 could be physical blocking of the p53K370me1 antibody by methylation at the neighbouring K372 residue. To test this possibility, we synthesized peptides bearing the double modification K370me1 and K372me1. We found that detection of K370me1 was comparable to that of the singly modified K370me1 peptide (Supplementary Fig. S10b).

We also found that higher concentrations of p53K370me1 occur when Set9 is decreased by siRNA in H1299 cells that ectopically express p53 (Fig. 4c). Finally, we found that lowering endogenous Set9 increases endogenous amounts of p53K370me1 in U2OS cells (Fig. 4d). On the basis of the *in vitro* and *in vivo* results, we conclude that Smyd2-mediated methylation of K370 is inhibited by Set9-mediated methylation of K372.

We considered that the inhibitory effect of p53K372me1 on p53K370me1 might be caused by blocking the interaction between p53 and Smyd2. We found that Flag–p53 coprecipitated Smyd2 in H1299 cells (Fig. 4e). Cotransfection of Set9 decreased the interaction

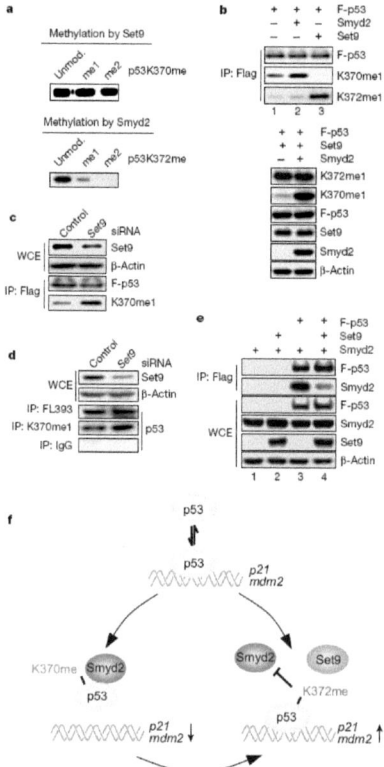

Figure 4 | Cross-talk between p53K370me1 and p53K372me1.
a, Autoradiogram of HMTase assays with recombinant Set9 or Smyd2 on p53 peptide 361–381. b, Western blot analysis of Flag immunoprecipitates (top) and whole-cell extract (bottom) from H1299 cells. c, Western blot analysis of whole-cell extract (WCE) and Flag immunoprecipitates from H1299 cells transfected with luciferase control or *Set9* siRNA. d, Western blot analysis of whole-cell extract and immunoprecipitates obtained with the indicated antibodies from U2OS cells. e, Western blot analysis of whole-cell extract and Flag immunoprecipitates from H1299 cells expressing Smyd2 either alone (lane 1) or with Flag–p53 (lane 3), Set9 (lane 2), or both Flag–p53 and Set9 (lane 4). f, Model of the mechanism of Smyd2-mediated repression of p53.

between p53 and Smyd2 (Fig. 4e, compare lanes 3 and 4). Therefore, Set9 prevents Smyd2 from binding to p53.

Our results suggest that there is an equilibrium between promoter-bound and free p53 (Fig. 4f): Smyd2-mediated methylation of K370 shifts the equilibrium towards dissociation of p53 from DNA, whereas Set9-mediated methylation of K372 enhances the association of p53 with promoter by blocking Smyd2-mediated methylation of K370 and promotes activation of the *p21* and *mdm2* genes.

Histone modification cross-talk occurs, such that methylation of histone H3 on K9 (H3-K9) precludes H3-K4 methylation or phosphorylation of histone H3 on S10 (H3-S10)[11,12]. We have shown here that the activating modification p53K372me1 causes interference to the repressing modification p53K370me1, which is reminiscent of the finding that Set9-mediated methylation of H3-K4 precludes Suv39h1-mediated methylation of H3-K9 (ref. 13). This non-proteolytic negative regulation through methylation of K370 may provide a pool of p53 protein to permit a quick response to DNA damage or cell stress.

Note added in proof: Smyd2 was recently shown to methylate histone H3 K36 (ref. 15).

METHODS

Histone methyltransferase assay. Histone methyltransferase (HMTase) assays were done as described[6]. In brief, 2 μg of peptide was incubated with 2 μg of recombinant enzyme and 1.1 μCi of *S*-adenosyl-methionine at 30 °C for 30 min.

Transfection. Cells were transfected with Lipofectamine 2000 (Invitrogen) for plasmid and with DharmaFECT 1 (Dharmacon) for siRNA (see Supplementary Information for details).

RT–PCR and real-time PCR. RT–PCR was done as described[14] (see Supplementary Information for details).

ChIP and sequential ChIP assays. We carried out ChIP assays as described[14]. Real-time PCR was used to measured signals in input material and immunoprecipitates. The percentage of input was calculated as the immunoprecipitate signal over the input signal. Sequential ChIP assays are described in the Supplementary Information.

Lentivirus transduction. Lentiviral RNAi and expression systems were purchased from Invitrogen. shRNA oligonucleotide sequences were designed with BLOCK-iT RNAi Designer and are given in the Supplementary Information. Lentivirus bearing p53 shRNA was purchased from Sigma.

Received 21 July; accepted 22 September 2006.
Published online 15 November 2006.

1. Zhang, Y. & Reinberg, D. Transcription regulation by histone methylation: interplay between different covalent modifications of the core histone tails. *Genes Dev.* 15, 2343–2360 (2001).
2. Lachner, M., O'Sullivan, R. J. & Jenuwein, T. An epigenetic road map for histone lysine methylation. *J. Cell Sci.* 116, 2117–2124 (2003).
3. Schotta, G. *et al.* A silencing pathway to induce H3-K9 and H4-K20 trimethylation at constitutive heterochromatin. *Genes Dev.* 18, 1251–1262 (2004).
4. Olivier, M., Hussain, S. P., Caron de Fromentel, C., Hainaut, P. & Harris, C. C. TP53 mutation spectra and load: a tool for generating hypotheses on the etiology of cancer. *IARC Sci. Publ.* 157, 247–270 (2004).
5. Vogelstein, B., Lane, D. & Levine, A. J. Surfing the p53 network. *Nature* 408, 307–310 (2000).
6. Levine, A. J. p53, the cellular gatekeeper for growth and division. *Cell* 88, 323–331 (1997).
7. Prives, C. & Hall, P. A. The p53 pathway. *J. Pathol.* 187, 112–126 (1999).
8. Chuikov, S. *et al.* Regulation of p53 activity through lysine methylation. *Nature* 432, 353–360 (2004).
9. Vaziri, H. & Benchimol, S. Reconstitution of telomerase activity in normal human cells leads to elongation of telomeres and extended replicative life span. *Curr. Biol.* 8, 279–282 (1998).
10. Luo, J. *et al.* Negative control of p53 by Sir2α promotes cell survival under stress. *Cell* 107, 137–148 (2001).
11. Wysocka, J., Myers, M. P., Laherty, C. D., Eisenman, R. N. & Herr, W. Human Sin3 deacetylase and trithorax-related Set1/Ash2 histone H3-K4 methyltransferase are tethered together selectively by the cell-proliferation factor HCF-1. *Genes Dev.* 17, 896–911 (2003).
12. Rea, S. *et al.* Regulation of chromatin structure by site-specific histone H3 methyltransferases. *Nature* 406, 593–599 (2000).
13. Nishioka, K. *et al.* Set9, a novel histone H3 methyltransferase that facilitates transcription by precluding histone tail modifications required for heterochromatin formation. *Genes Dev.* 16, 479–489 (2002).
14. Kent, J. R. *et al.* During lytic infection herpes simplex virus type 1 is associated with histones bearing modifications that correlate with active transcription. *J. Virol.* 78, 10178–10186 (2004).
15. Brown, M. A., Sims, R. J. III, Gottlieb, P. D. & Tucker, P. W. Identification and characterization of Smyd2: a split SET/MYND domain-containing histone H3 lysine 36-specific methyltransferase that interacts with the Sin3 histone deacetylase complex. *Mol. Cancer* 5, 26 (2006).

Supplementary Information is linked to the online version of the paper at www.nature.com/nature.

Acknowledgements We thank N. Barlev for the Set9 expression vector and D. Reinberg for the p53K372me1 antibody; T. Waibel for assistance in cloning Smyd2; S. Benchimol for the BJ and BJ-DNp53 cell lines; and members of the T.J. and S.L.B. laboratories for discussions. Research in the laboratory of T.J. is supported by the IMP through Boehringer Ingelheim and by grants from the European Union and the Austrian GEN-AU initiative, which is financed by the Austrian Ministry of Education, Science and Culture. Research support to S.L.B. was provided by a grant from the NIH. B.J.P. was supported by a Wistar Cancer Training Grant.

Author Information Reprints and permissions information is available at www.nature.com/reprints. The authors declare no competing financial interests. Correspondence and requests for materials should be addressed to S.L.B. (berger@wistar.org).

Supplementary information:

Methods
Constructs and antibodies

Mouse Smyd2, Smyd3, Smyd5 and Suv4-20h1 were cloned into pGEX vectors and purified from BL21 cells. cDNAs of mouse Smyd2 and Smyd2(MD) were subcloned into mammalian expression vector with or without FLAG tag. Antibodies specific to p53K370me1, me2 and me3 were raised in rabbit, using p53 peptides as antigens, and purified as IgG fraction. Smyd2 antibody was raised against amino terminal peptide (NH2-CKDHPYISEIKQEIESH-COOH) and antigen purified from crude serum.

RT-PCR and realtime PCR

Real-time PCR was run on an ABI 7000 SDS machine. mRNA Levels were measured by real-time PCR and normalized to that of GAPDH. The relative mRNA level was calculated by comparing the normalized values to that at 0-hour time point treated with control siRNA, the value of which was set to 1. Primers for GAPDH: 5'-TGGGCTACACTGAGCACCAG-3' and 5'-GGGTGTCGCTGTTGAAGTCA-3'; for p21: 5'-AGCGATGGAACTTCGACTTTG-3' and 5'-CGAAGTCACCCTCCAGTGGT-3'; for mdm2: 5'-CCGGATCTTGATGCTGGTGT-3' and 5'-CTGATCCAACCAATCACCTGAAT-3'

siRNA transfection

For siRNA transfection, cells were transfected with two rounds of 100nM siRNA (Dharmacon) and DharmaFECT 1 (Dharmacon) with 24-hour interval. Cells were treated with or without 0.5uM adriamycin as indicated. siRNA target sequence for:
luciferase control: 5'-UAAGGCUAUGAAGAGAUAC-3';
for Smyd2: 5'-GCAAAGAUCAUCCAUAUAU-3';
for Set9: 5'- UGUAGACGGAGAGCUGAAC-3'.

Re-ChIP (Sequential ChIP)

H1299 cells were transfected with FLAG-p53 or FLAG-p53(K370R) together with or without Smyd2 or Smyd2(MD) expression vector. After the first ChIP with FLAG antibody, IPed materials were eluted with FLAG peptide and subject to Re-ChIP with p53K370me1 antibody or IgG. Percentage of input was calculated as IP signal (Re-ChIP)/IP signal (ChIP). Primers for p53 binding site
on p21: 5'-GGCTGGTGGCTATTTTGTCC-3' and 5'-TCCCCTTCCTCCCTGAAAAC-3';
on mdm2: 5'-AAACCATGCATTTTCCCAGC-3' and 5'-CAGGTCTACCCTCCAATCGC-3';
on Bax: 5'-AGCGTTCCCCTAGCCTCTTT-3' and 5'-GCTGGGCCTGTATCCTACATTCT-3'.

Lentivirus transduction

Luciferase control
Top strand:
5'CACCGCCCTGGTTCCTGGAACAATTCGAAAATTGTTCCAGGAACCAGGGC3'
Bottom strand:
5'AAAAGCCCTGGTTCCTGGAACAATTTTCGAATTGTTCCAGGAACCAGGGC3'
Smyd2
Top strand:
5'CACCGGATTGTCCAAATGTGGAAGACGAATCTTCCACATTTGGACAATCC3'
Bottom strand:
5'AAAAGGATTGTCCAAATGTGGAAGATTCGTCTTCCACATTTGGACAATCC3'

Lentiviruses were produced per manufacturer's protocol. Stable clones constitutively expressing shRNAs were obtained by transducing cells with lentivirus expressing shRNAs followed by selecting with 250ug/ml (U2OS) or 100ug/ml (H1299) Zeocin for two weeks. For inducible stable clones, U2OS cells expressing TetR were obtained by transducing with Lenti6/TR lentivirus and selecting with 2.5ul/ml Blasticidin for one week. U2OS/Lenti6/TR cells were then transduced with Lentivirus expressing FLAG-Smyd2 and selected with Zeocin. Cells transduced with p53 shRNA lentivirus were selected with 1ug/ml puromycin and 40ug/ml Zeocin for three days.

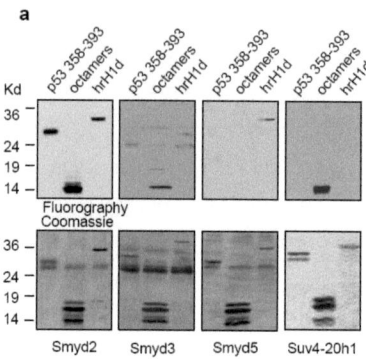

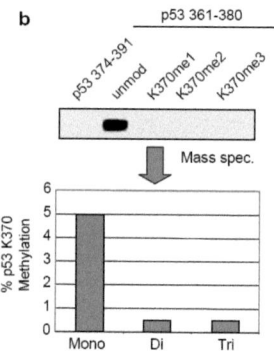

Figure S1. a) Autoradiograms and Coomassie stainings of histone methyltransferase (HMTase) assays with recombinant Smyd2, Smyd3, Smyd5 and Suv4-20h1 on p53 peptide 358-393, histone octamers or linker histone H1d. **b)** Mass spectrometry analysis of indicated p53 peptides methylated by Smyd2.

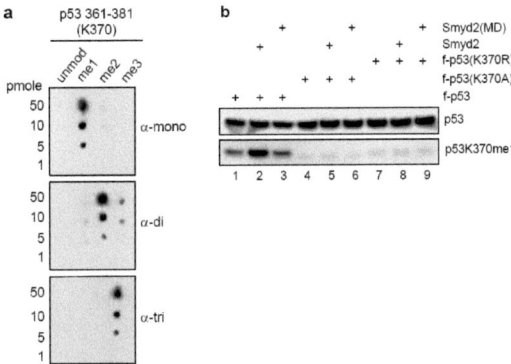

Figure S2. a) Dotblot analysis with antibodies raised against unmodified, mono-, di- or tri-methylation of p53K370 on peptides corresponding to p53 361-381. **b)** Detection of p53K370me1 by western blot. Expression vector encoding FLAG-p53 (lane1-3), FLAG-p53(K370A) (lane4-6) or FLAG-p53(K370R) (lane7-9) were co-transfected with empty vector (lane1, 4, 7) or expression vector for Smyd2 (lane 2, 5, 8) or for Smyd2(MD) (lane3, 6, 9). Eluates from FLAG immunoprecipitation were subjected to western blot analysis with DO1 and p53K370me1 antibodies to detect the total p53 and p53K370 methylation, respectively.

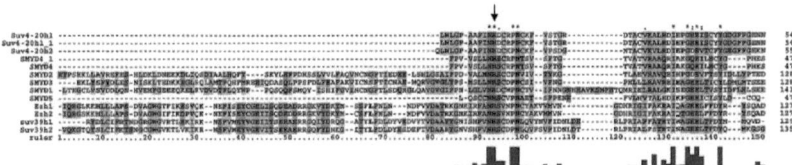

Figure S3. Protein sequence alignment of SET domains of Set domain containing proteins. Highly conserved residues are marked by asterisks. The Histidine 207 (arrow) in Smyd2 was mutagenized to Alanine to generate methylation defective (MD) mutant of Smyd2.

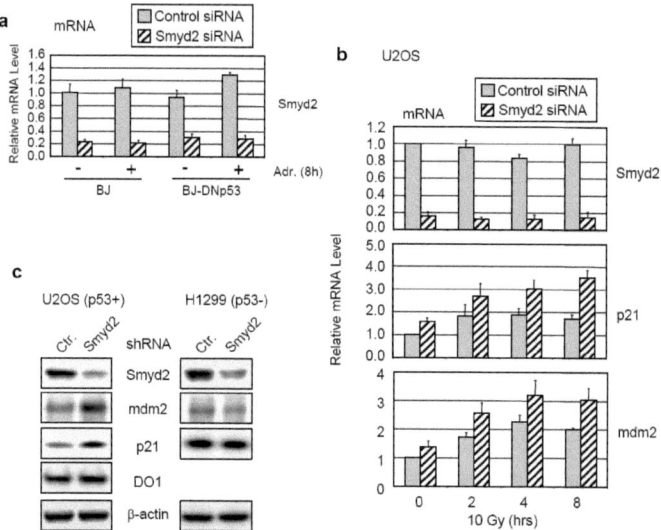

Figure S4. Smyd2 regulates p53 responsive genes. **a)** Realtime PCR analysis of relative mRNA level of Smyd2 from BJ and BJ-DNp53 cells transfected with control (grey bar) or Smyd2 (striped bar) siRNA followed by adriamycin treatment for 8 hours. Results were the average from two independent experiments measured as triplicates. **b)** Realtime PCR analysis of relative mRNA level of Smyd2, p21 and mdm2 from U2OS cells transfected with control (grey bar) or Smyd2 (striped bar) siRNA followed by adriamycin treatment for indicated time. Results were the average from two independent experiments measured as triplicates. **c)** Western blot analysis to detect protein levels of Smyd2, p21, mdm2, DO1 and β-actin in U2OS and H1299 cells stably expressing shRNA against luciferase (Ctr.) or Smyd2 mRNA.

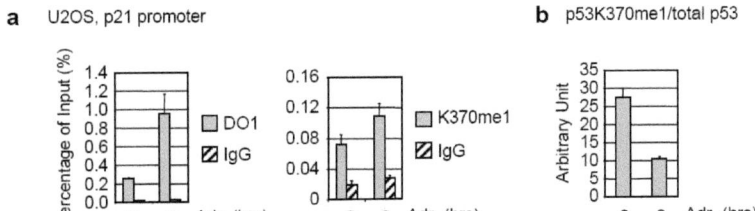

Figure S5. ChIP assay to measure the promoter associated total p53 and methylated p53 at K370. **a)** ChIP assay with DO1 and p53K370me1 antibody in U2OS treated with adriamycin for indicated time followed by realtime PCR analysis. **b)** relative percentage of methylated p53 at K370 bound to p21 promoter was calculated by dividing p53K370me1 signal by total p53 signal.

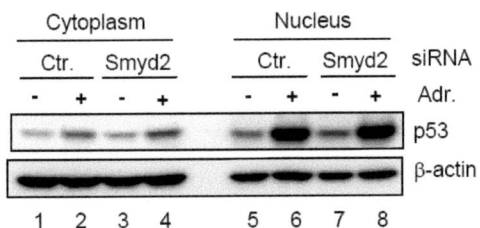

Figure S6. Western analysis to measure effect of Smyd2 on the distribution of p53 between cytoplasm and nucleus. U2OS cells were transfected with control (lane 1, 2, 5 and 6) or Smyd2 (lane 3, 4, 7 and 8) siRNA followed by mock (lane 1, 3, 5 and 7) or adriamycin (lane 2, 4, 6 and 8) treatment for 8 hours. 25ug cytoplasmic (lane 1-4) or nuclear lysate (lane 5-8) was loaded for each lane followed by western blot analysis with p53 specific antibody, DO1.

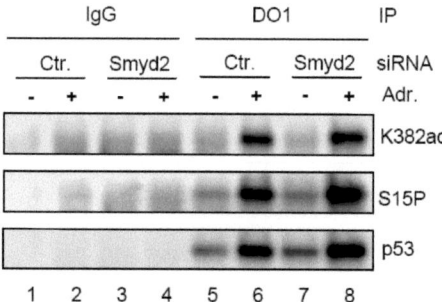

Figure S7. Western analysis to measure effect of Smyd2 on the level of p53K382 acetylation and p53S15 phosphorylation. U2OS cells were transfected with control (lane 1, 2, 5 and 6) or Smyd2 (lane 3, 4, 7 and 8) siRNA followed by mock (lane 1, 3, 5 and 7) or adriamycin (lane 2, 4, 6 and 8) treatment for 8 hours. Lysates were immunoprecipitated with IgG (lane 1-4) or DO1 (lane 5-8) and followed by western blot analysis with p53K382 acetylation (top), p53S15 phosphorylation (middle) and p53 (bottom, FL393) specific antibodies.

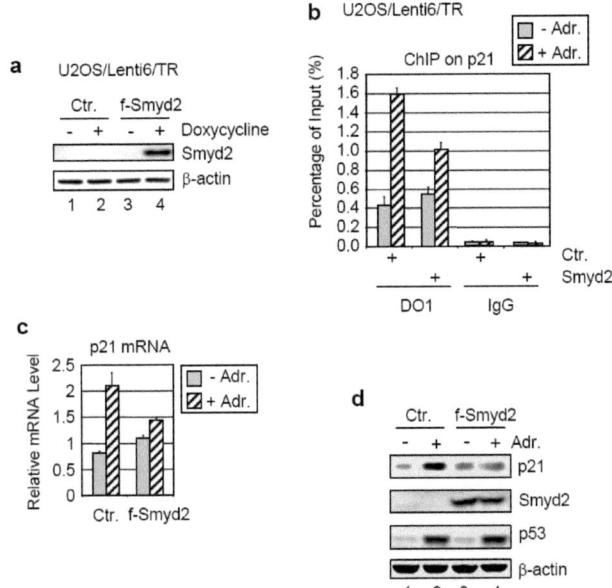

Figure S8. Effect of Smyd2 overexpression on the levels of p21 promoter-associated p53. **a)** Western blot analysis of Smyd2 in lysates from U2OS/Lenti6/TR cells inducibly expressing vector (Ctr.) (lanes 1 and 2) or FLAG-Smyd2 (f-Smyd2) (lanes 3 and 4) treated without (lanes 1 and 3) or with (lanes 2 and 4) doxycycline (Dox.) for 24 hours. **b)** ChIP assay of U2OS cell lines inducibly expressing none or FLAG-Smyd2 with DO1 or IgG after 8h adriamycin treatment. **c)** the relative mRNA level of p21 in U2OS/lenti6/TR cells expressing vector or f-Smyd2 treated without (grey bar) or with (striped bar) adriamycin for 8 hours. **d)** Western blot analysis of vector control (lanes 1 and 2) or f-Smyd2 (lanes 3 and 4) cells incubated with Doxycyline for 24 hours followed by mock (lanes 1 and 3) or adriamycin treatment (lanes 2 and 4) for 8 hours.

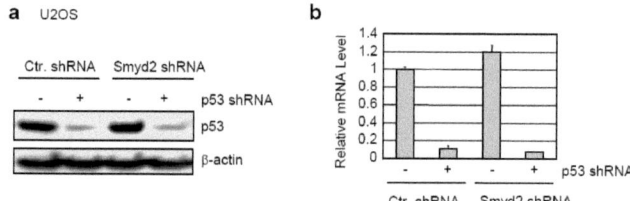

Figure S9. Measurement of p53 protein and mRNA levels. U2OS cells stably expressing control or Smyd2 shRNA (levels shown in S4c) were transduced with lentivirus carrying a luciferase control or p53 shRNA. Western blot analysis of the protein with DO1 **a)** and realtime PCR analysis of the mRNA **b)** of p53.

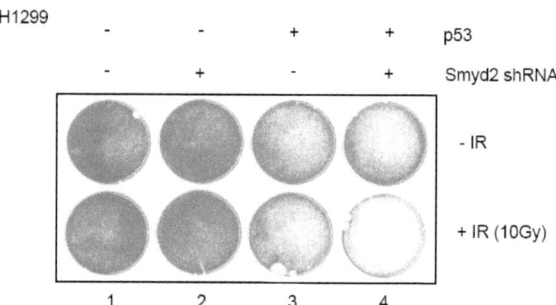

Figure S10. Effect of Smyd2 on p53-mediated apoptosis. H1299/Ctr shRNA (column 1 and 3) and H1299/Smyd2 shRNA (column 2 and 4) cells were transfected with empty (column 1 and 2) or p53 (column 3 and 4) expression vectors followed by mock (top) or 10Gy gamma irradiation (bottom) for 12 hours. Apoptotic cells were washed off and plates were stained by crystal violet. The level of H1299/Ctr shRNA and H1299/Smyd2 shRNA are shown in Fig. S4c.

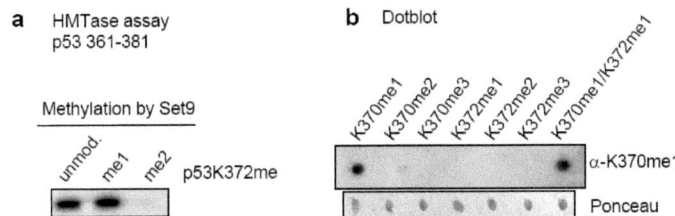

Figure S11. a) Set9 methylation of p53 peptides in vitro. Autoradiograph of HMTase assay with recombinant Set9 and p53 peptide un-, mono- or di-methylated at K372. **b)** K372me1 does not interfere with the recognition of p53K370me1 antibody on K370me1. Dotblot analysis with p53K370me1 antibody and Ponceau staining.

4.3. p53 is regulated by the lysine demethylase LSD1

Jing Huang[1], Roopsha Sengupta[2], Alexsandra B. Espejo[3], Min Gyu Lee[1], Jean A. Dorsey[1], Mario Richter[2], Susanne Opravil[2], Ramin Shiekhattar[1], Mark T. Bedford[3], Thomas Jenuwein[2] & Shelley L. Berger[1]

p53, the tumour suppressor and transcriptional activator, is regulated by numerous post-translational modifications, including lysine methylation[1,2]. Histone lysine methylation has recently been shown to be reversible; however, it is not known whether non-histone proteins are substrates for demethylation. Here we show that, in human cells, the histone lysine-specific demethylase LSD1 (refs 3, 4) interacts with p53 to repress p53-mediated transcriptional activation and to inhibit the role of p53 in promoting apoptosis. We find that, in vitro, LSD1 removes both monomethylation (K370me1) and dimethylation (K370me2) at K370, a previously identified Smyd2-dependent monomethylation site[2]. However, in vivo, LSD1 shows a strong preference to reverse K370me2, which is performed by a distinct, but unknown, methyltransferase. Our results indicate that K370me2 has a different role in regulating p53 from that of K370me1: K370me1 represses p53 function, whereas K370me2 promotes association with the coactivator 53BP1 (p53-binding protein 1) through tandem Tudor domains in 53BP1. Further, LSD1 represses p53 function through the inhibition of interaction of p53 with 53BP1. These observations show that p53 is dynamically regulated by lysine methylation and demethylation and that the methylation status at a single lysine residue confers distinct regulatory output. Lysine methylation therefore provides similar regulatory complexity for non-histone proteins and for histones.

Previous observations of p53 methylation led us to test whether LSD1, also known as BHC110 (ref. 3), has a role in p53 signalling. We used co-immunoprecipitation assays to examine whether LSD1 binds to p53. Ectopically expressed tagged LSD1 and p53 associated in HEK-293 cells (Fig. 1a). HDAC2, which is present in a stable complex with LSD1 (ref. 5), also co-immunoprecipitated with p53 (Fig. 1a). Using nuclear extract, we detected interaction between endogenous LSD1 and p53 in MCF7 cells with or without treatment by adriamycin, a DNA-damaging reagent that activates p53 (Fig. 1b). Similar results were obtained in HEK-293 cells (Supplementary Fig. 1).

We studied whether LSD1 interacts directly with p53. Purified recombinant Flag-LSD1 from insect cells (Supplementary Fig. 3a) and glutathione S-transferase (GST)-tagged p53 from bacteria were used in GST pull-down assays. We observed that GST–p53, but not GST alone, associates with Flag–LSD1, suggesting a direct interaction between p53 and LSD1 (Fig. 1c).

LSD1 can act either as a transcriptional coactivator or as a co-repressor[4,6]. We investigated the role of LSD1 in p53-mediated transcriptional regulation in either mock-treated or adriamycin-treated U2OS cells. Ablation of LSD1 with short interfering RNA (siRNA) results in increased expression of p21 and mdm2 (Fig. 2a), two well-characterized p53 target genes. The increased expression of p21 and mdm2 was not caused by an increased steady-state level of p53 after treatment with LSD1 siRNA (Supplementary Fig. 2a). In fact, we consistently observed a slight decrease in the steady-state level of p53 without adriamycin treatment, which probably resulted from feedback proteolysis of p53 caused by an elevated level of mdm2 when that of LSD1 was decreased. Similar results were observed in MCF7 cells (Supplementary Fig. 2b, c). To examine whether the effect of LSD1 knockdown on the expression of p21 and mdm2 is p53-dependent, we used the cell lines BJ and BJ-DNp53 (refs 2, 7). BJ-DNp53 cells contain a stably incorporated dominant-negative p53 (DNp53) that prevents endogenous p53 from binding to DNA[7,8]. We observed that a decrease in LSD1 level increased the expression of p21 in BJ cells, but not in BJ-DNp53 cells (Fig. 2b and Supplementary Fig. 2d). A similar dependence of LSD1-mediated repression on p53 was observed in another pair of cell lines, HCT116 ($p53^{+/+}$) and HCT116 ($p53^{-/-}$) (Supplementary Fig. 2e). Taken together, these results show that LSD1 represses the transcriptional activity of p53.

Because LSD1 does not regulate the activity of p53 in BJ-DNp53 cells in which the DNA-binding ability of p53 is impaired, we reasoned that LSD1 might affect the binding of DNA by p53, thus

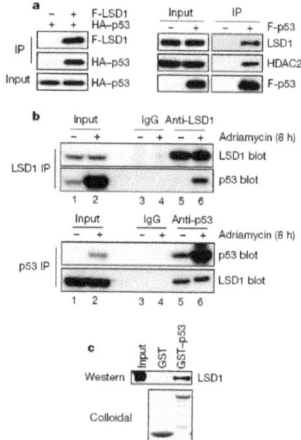

Figure 1 | LSD1 interacts with p53. a, b, Reciprocal co-immunoprecipitation assays were used to detect interaction between ectopic (a) and endogenous (b) LSD1 and p53 in HEK-293 cells (a) and MCF7 cells (b). **c,** GST pull-down assay. F, Flag; HA, haemagglutinin; IP, immunoprecipitation.

[1]The Wistar Institute, 3601 Spruce Street, Philadelphia, Pennsylvania 19104, USA. [2]Research Institute of Molecular Pathology (IMP), The Vienna Biocenter, 1030 Vienna, Austria. [3]M. D. Anderson Cancer Center, Department of Carcinogenesis, University of Texas, Smithville, Texas 78957, USA.

altering the equilibrium between DNA-bound and free p53. To test this hypothesis, we used chromatin immunoprecipitation (ChIP) and real-time polymerase chain reaction (PCR). We found significantly higher levels of p53 at the *p21* promoter in U2OS cells than in control cells when the level of LSD1 was decreased by siRNA treatment (Fig. 2c). These results indicate that LSD1 represses the activity of p53 in part through decreasing the binding of p53 to DNA.

After DNA damage, p53 activates the expression of genes that are involved in either cell cycle arrest or apoptosis. Because LSD1 represses the transcriptional activation activity of p53, we proposed that LSD1 inhibits p53-mediated apoptosis and/or cell cycle arrest. Without adriamycin treatment, decreasing LSD1 levels with lentivirus-based short hairpin RNA (shRNA) did not affect the percentage of apoptotic cells (sub-G1 population) in U2OS cells (Fig. 2d, upper panels, and Supplementary Fig. 2f). However, the apoptotic cell population increased (to 19%) in response to adriamycin treatment, and increased further (to 33.3%) when LSD1 levels were decreased with shRNA (Fig. 2d, lower panels). These results indicate that LSD1 represses p53-mediated apoptosis.

LSD1 was the first lysine demethylase to be characterized; it demethylates lysine 4 or lysine 9 on histone H3 (refs 4–6). Because LSD1 binds directly to p53 (Fig. 1c), we examined whether LSD1 demethylates p53. To assay the two known p53 methylation sites *in vitro*, we developed a methylation–demethylation assay (Supplementary Fig. 3a–c and Methods). We found that LSD1 specifically removes Smyd2-mediated methylation but not methylation catalysed by Set9 (Fig. 3a). Because K370 is the only amino acid residue modified by Smyd2 *in vitro* (Supplementary Fig. 3d), we conclude that LSD1 demethylates p53 at K370 *in vitro*.

Smyd2 is a monomethyl transferase[2], and LSD1 is able to remove both monomethylation and dimethylation from histone substrates[4]. To determine whether LSD1 can remove dimethylation at p53K370, we used peptides representing monomethylated (K370me1), dimethylated (K370me2) and trimethylated (K370me3) substrate, as well as monomethylated (K372me1) p53K372 (Fig. 3b). Reaction products were subjected to dot-blot analysis to assess the different methylation sites and levels of methylation by using specific antibodies[1,2]. We found that LSD1 decreases the level of K370me1 and K370me2 but not that of K370me3 (Fig. 3b), supporting previous findings that LSD1 demethylates only monomethylated and dimethylated lysine residues on histone substrates[4]. Consistent with the observation on methylated GST–p53 (Fig. 3a) was our observation that LSD1 was not able to demethylate K372me1 peptide (Fig. 3b). These results show that LSD1 specifically removes both K370me1 and K370me2 *in vitro*.

We next investigated whether LSD1 affects K370 or K372 methylation *in vivo*. The level of LSD1 was decreased in U2OS cells by using siRNA, and then K370me1, K370me2 and K372me1 signals were determined by western blot analysis after immunoprecipitation with p53 antibody. On adriamycin treatment, a decrease in LSD1 level significantly increased that of K370me2, whereas that of K370me1

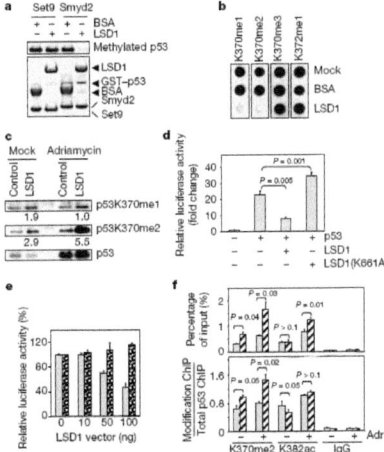

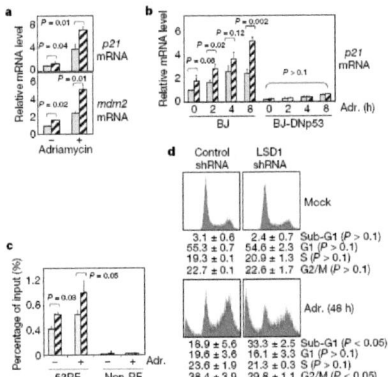

Figure 2 | LSD1 represses the activity of p53. a, Quantitative real-time PCR to assess the relative mRNA levels of *p21* and *mdm2* in U2OS cells transfected with control (grey columns) or LSD1 (hatched columns) siRNA followed by 24 h of mock treatment or treatment with adriamycin. **b**, Quantitative real-time PCR was used to measure the relative mRNA levels of p21 in BJ and BJ-DNp53 cells transfected with control (grey columns) or LSD1 (hatched columns) siRNA followed by treatment with adriamycin (Adr.) for 0, 2, 4 or 8 h. **c**, ChIP assay to measure p53 bound to its response element (53RE) and to a control site (non-RE) within the *p21* promoter in U2OS cells transfected with control (grey columns) or LSD1 (hatched columns) siRNA followed by treatment with adriamycin for 8 h. **d**, Flow cytometry analysis to assess the effect of LSD1 on the percentage of apoptotic (that is, sub-G1) U2OS cells. Results are shown as means ± s.d.

Figure 3 | LSD1 demethylates p53 at K370. a, Top: fluorography, showing demethylation of p53 by LSD1. Bottom: colloidal staining, showing loadings. **b**, Dot-blot analysis, showing demethylation of p53 peptides by LSD1. **c**, Western blot analysis of DO1 immunoprecipitation in U2OS cells transfected with control or LSD1 siRNA followed by treatment with adriamycin for 8 h. Numbers below the western blots show the ratio of the signal with LSD1 knockdown to control knockdown, then to p53 levels. **d, e**, Dual luciferase assay in H1299 cells with a reporter bearing *p21* promoter to show that LSD1-mediated repression is dependent on its demethylation activity (**d**) and K370 in p53 (**e**). In **e**: grey columns, wild-type p53; stippled columns, p53(K370R). **f**, ChIP assay to detect p53K370me2 and p53K382ac recruitment to *p21* promoter in U2OS cells transfected with control (grey columns) or LSD1 (hatched columns) siRNA followed by treatment with adriamycin (Adr.) for 8 h. Top: absolute percentage of input. Bottom: ratio of modification ChIP to total p53 ChIP. Results in **d–f** are shown as means ± s.d.

was affected only slightly (Fig. 3c). Therefore, although LSD1 is able to remove both K370me1 and K370me2 *in vitro*, it seems to demethylate K370me2 preferentially *in vivo*. Other histone demethylases also exhibit a preference for specific methylation levels[8,10]. The underlying mechanism for these preferences is unclear but may involve other post-translational modifications on the substrate or demethylase-associated proteins. Under these assay conditions, endogenous K370me3 and K372me1 signals are not detected, even after long exposures in western blot analysis (data not shown). We note that Smyd2 is a monomethyl transferase[2]; our results therefore indicate that K370me2 is performed by a distinct, currently unidentified methyltransferase. Taken together, our results reveal that p53K370me2 is demethylated by LSD1.

To assess whether the demethylation activity of LSD1 is required for its repression of the transcriptional activity of p53, we ectopically expressed *LSD1* and *LSD1(K661A)*, a demethylation-defective mutant[5], together with p53 in H1299 cells. A luciferase reporter bearing the *p21* promoter was used to assess transcriptional activation driven by p53 (Fig. 3d). We observed that LSD1 but not LSD1(K661A) decreases the activity of p53 (Fig. 3d). LSD1 and LSD1(K661A) interact with p53 comparably (Supplementary Fig. 3e), suggesting that LSD1 represses the activity of p53 in a demethylation-dependent manner. To test whether LSD1 represses the activity of p53 in a K370-dependent manner, we co-transfected H1299 cells with vectors expressing *LSD1* along with either *p53* or *p53(K370R)*. We found that LSD1 decreased luciferase activity mediated by wild-type p53 but not by p53(K370R) (Fig. 3e), demonstrating that LSD1 represses the activity of p53 through K370.

We determined whether p53 dimethylated at K370 is present at the *p21* promoter in U2OS cells by ChIP assay with DO1 (unmodified p53), p53K370me2 and p53K382 acetylation (K382ac) antibodies (Fig. 3f). The K382ac antibody serves as a positive control because it detects a well-characterized activating modification of p53 during DNA damage[11,12]. Treatment with adriamycin results in increased K370me2 and K382ac ChIP signals (Fig. 3f, grey bars), suggesting that K370me2 is an activating modification for p53. Decreasing the level of LSD1 by siRNA results in a strongly increased K370me2 signal, whereas the K382ac signal is enhanced only slightly (Fig. 3f, upper panel). The K370me2 level increases even after normalization to the total p53 level (Fig. 3f, lower panel), showing that higher K370me2 is due not only to elevated p53 levels at the promoter caused by treatment with LSD1 siRNA (Fig. 2c). Thus, p53K370me2 increases at the *p21* promoter after DNA damage and this recruitment is regulated by LSD1.

These results suggest that K370me2 may have a distinct role in regulating p53 in comparison with repression-linked Smyd2-mediated K370me1. Recent studies show that specific lysine methylation states in histones serve as recognition sites for specific binding proteins[13–17]. We investigated binding proteins that recognize specific methylation states at K370 by screening a GST protein domain microarray[18] (Fig. 4a and Supplementary Fig. 4). We found that the tandem Tudor domains of 53BP1 preferentially recognize K370me2 peptide, compared with K370me0, K370me1 or K370me3 (Fig. 4a). We used peptide pull-down to confirm that 53BP1 binds to the K370me2 peptide more strongly than to K370me1 or K370me3 peptide, and that there is no binding to the K370me0 peptide (Supplementary Fig. 5a). These results are similar to recent observations *in vitro* that the tandem Tudor domains of 53BP1 preferentially bind to histone dimethylated at H3K79 and H4K20, and bind more weakly to monomethylated H3K79 and H4K20 (refs 18–20).

We next examined whether 53BP1 associates with dimethylated p53K370 more strongly than with monomethylated p53K370 *in vivo*. We performed a Flag immunoprecipitation assay to pull down Flag–53BP1-associated proteins followed by western blot analysis to detect p53K370me1 and K370me2 (Fig. 4b). Because the p53K370me1 signal is weaker than that of K370me2, Smyd2 was coexpressed with p53 to provide equivalent K370me1 input. Under the same loading conditions of input signals, we detected more K370me2 signal than K370me1 in the Flag–53BP1 immunoprecipitation eluates (Fig. 4b). These results *in vivo* support the hypothesis that 53BP1 binds to p53K370me2 more strongly than to p53K370me1.

Physical interaction between p53 and 53BP1 was discovered in a yeast two-hybrid analysis[21], and *in vivo* interaction between p53 and 53BP1 has recently been reported[22,23]. We are able reproducibly to detect interaction between endogenous p53 and 53BP1 *in vivo* (Supplementary Fig. 5b). The specificity of this interaction

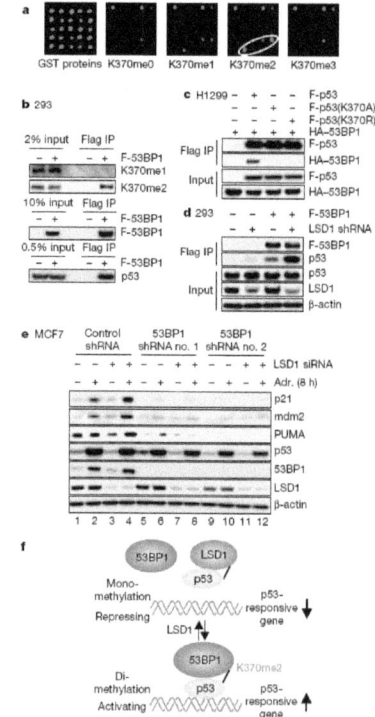

Figure 4 | LSD1 represses the activity of p53 through 53BP1. a, GST protein domain array to identify binding proteins for p53K370me0, p53K370me1, p53K370me2 and p53K370me3. The K370me2-dependent interaction with tandem Tudor domains of 53BP1 is circled. Part of the protein array (C3 of Tudor domain section) is shown. For the entire array, see Supplementary Fig. 4. **b**, Flag immunoprecipitation (IP) to assay the relative binding of 53BP1 to K370me1 and K370me2 *in vivo*. **c**, Flag IP to determine the requirement of K370 for the interaction between 53BP1 and p53. **d**, Flag IP to study the effect of LSD1 knockdown on the 53BP1–p53 interaction. **e**, Western blot analysis to test whether LSD1-mediated repression is 53BP1-dependent. Adr., adriamycin. **f**, Schematic model for the dynamic regulation of K370 methylation.

was analysed in several approaches. First, we generated 53BP1 bearing substitution mutations in conserved residues in the Tudor domains previously shown to be required for binding to histone H3K79me2 and H4K20me2 (Y1502Q or D1521R)[19,20]. These substitutions in 53BP1 strongly decreased co-immunoprecipitation with p53 (Supplementary Fig. 5c). Second, p53 substituted at K370 (K370A or K370R) showed greatly decreased interaction with 53BP1 (Fig. 4c). Last, decreasing the level of LSD1 with shRNA increased the interaction between p53 and 53BP1 (Fig. 4d), indicating that LSD1 represses the activity of p53 through decreasing the interaction with 53BP1.

53BP1 has been functionally linked to p53 as a potential coactivator[21,24,25]. We directly assayed whether 53BP1 is required for transcriptional activation by p53. 53BP1 levels in MCF7 cells were stably reduced by lentivirus-based shRNAs (Fig. 4e). Two different shRNAs targeting 53BP1 for knockdown resulted in decreased protein levels of p21, mdm2 and PUMA (p53 upregulated modulator of apoptosis) (Fig. 4e, compare lane 2 with lane 6 or 10) and messenger RNA (Supplementary Fig. 5d). As described above (Fig. 2 and Supplementary Fig. 2), treatment with LSD1 siRNA increased the expression of *p21*, *mdm2* and *PUMA* in MCF7 cells (Fig. 4e, compare lanes 2 and 4). In contrast, gene expression did not increase in the LSD1 knockdown in cells also carrying 53BP1 shRNAs (Fig. 4e, compare lanes 2 and 4 with lanes 6 and 8 or 10 and 12). Together, these results strongly indicate that LSD1 regulates the activity of p53 through its coactivator 53BP1.

Our results demonstrate that LSD1 demethylates p53 at K370 and displays a strong preference *in vivo* for demethylating p53K370me2. In addition, although we did not detect K372me1 as a substrate, previous observation that K372me1 is transient during p53 activation[1] indicates that other demethylases might target methylated p53.

It is well established in histones that various methylation levels elicit specific biological outcomes, through recognition by different binding proteins[13–15]. In this study we found that K370me2, in contrast to K370me1, has an activating role in p53 regulation through providing an interaction surface for the binding of 53BP1. These results add another layer of complexity to the methylation-mediated regulation of p53. Methylation, similarly to other post-translational modifications of p53, seems to fine-tune p53 function[26].

We propose a dynamic model for the regulation of p53 through lysine methylation (Fig. 4f). During gene activation, p53, by means of K370me2, binds to 53BP1. In contrast, during gene repression, LSD1 prevents the accumulation of K370me2 by demethylating the site, thereby disallowing the binding of 53BP1 to p53. The demethylation activity of LSD1 thus maintains p53 in an inactive state, to prevent binding to DNA. 53BP1 and p53 synergize to inhibit tumorigenesis[27]; our results suggest a mechanism contributing to their cooperative suppression of cancer.

METHODS SUMMARY

Co-immunoprecipitation and GST pull-down assays were used to detect protein–protein interaction. Transient transfection of siRNA and lentivirus transduction of shRNA were used to decrease the levels of specific proteins. Flow cytometry was used to assess the apoptosis and cell cycle arrest caused by p53. ChIP assay was performed to detect the recruitment of p53 and modified p53 to DNA. We used real-time PCR to measure the mRNA level and immunoprecipitated DNA in a ChIP assay. An *in vitro* demethylation assay was used to determine whether LSD1 demethylates p53. A protein domain microarray was used to identify proteins that bind to p53K370me1, p53K370me2 or p53K370me3.

Full Methods and any associated references are available in the online version of the paper at www.nature.com/nature.

Received 5 April; accepted 9 July 2007.

1. Chuikov, S. et al. Regulation of p53 activity through lysine methylation. *Nature* 432, 353–360 (2004).
2. Huang, J. et al. Repression of p53 activity by Smyd2-mediated methylation. *Nature* 444, 629–632 (2006).
3. Hakimi, M. A., Dong, Y., Lane, W. S., Speicher, D. W. & Shiekhattar, R. A candidate X-linked mental retardation gene is a component of a new family of histone deacetylase-containing complexes. *J. Biol. Chem.* 278, 7234–7239 (2003).
4. Shi, Y. et al. Histone demethylation mediated by the nuclear amine oxidase homolog LSD1. *Cell* 119, 941–953 (2004).
5. Lee, M. G., Wynder, C., Cooch, N. & Shiekhattar, R. An essential role for CoREST in nucleosomal histone 3 lysine 4 demethylation. *Nature* 437, 432–435 (2005).
6. Metzger, E. et al. LSD1 demethylates repressive histone marks to promote androgen-receptor-dependent transcription. *Nature* 437, 436–439 (2005).
7. Vaziri, H. & Benchimol, S. Reconstitution of telomerase activity in normal human cells leads to elongation of telomeres and extended replicative life span. *Curr. Biol.* 8, 279–282 (1998).
8. Shaulian, E., Zauberman, A., Ginsberg, D. & Oren, M. Identification of a minimal transforming domain of p53: negative dominance through abrogation of sequence-specific DNA binding. *Mol. Cell. Biol.* 12, 5581–5592 (1992).
9. Tsukada, Y. et al. Histone demethylation by a family of JmjC domain-containing proteins. *Nature* 439, 811–816 (2006).
10. Yamane, K. et al. JHDM2A, a JmjC-containing H3K9 demethylase, facilitates transcription activation by androgen receptor. *Cell* 125, 483–495 (2006).
11. Barlev, N. A. et al. Acetylation of p53 activates transcription through recruitment of coactivators/histone acetyltransferases. *Mol. Cell* 8, 1243–1254 (2001).
12. Gu, W. & Roeder, R. G. Activation of p53 sequence-specific DNA binding by acetylation of the p53 C-terminal domain. *Cell* 90, 595–606 (1997).
13. Wysocka, J. et al. WDR5 associates with histone H3 methylated at K4 and is essential for H3 K4 methylation and vertebrate development. *Cell* 121, 859–872 (2005).
14. Wysocka, J. et al. A PHD finger of NURF couples histone H3 lysine 4 trimethylation with chromatin remodelling. *Nature* 442, 86–90 (2006).
15. Shi, X. et al. ING2 PHD domain links histone H3 lysine 4 methylation to active gene repression. *Nature* 442, 96–99 (2006).
16. Pena, P. V. et al. Molecular mechanism of histone H3K4me3 recognition by plant homeodomain of ING2. *Nature* 442, 100–103 (2006).
17. Li, H. et al. Molecular basis for site-specific read-out of histone H3K4me3 by the BPTF PHD finger of NURF. *Nature* 442, 91–95 (2006).
18. Kim, J. et al. Tudor, MBT and chromo domains gauge the degree of lysine methylation. *EMBO Rep.* 7, 397–403 (2006).
19. Huyen, Y. et al. Methylated lysine 79 of histone H3 targets 53BP1 to DNA double-strand breaks. *Nature* 432, 406–411 (2004).
20. Botuyan, M. V. et al. Structural basis for the methylation state-specific recognition of histone H4-K20 by 53BP1 and Crb2 in DNA repair. *Cell* 127, 1361–1373 (2006).
21. Iwabuchi, K., Bartel, P. L., Li, B., Marraccino, R. & Fields, S. Two cellular proteins that bind to wild-type but not mutant p53. *Proc. Natl Acad. Sci. USA* 91, 6098–6102 (1994).
22. Sengupta, S. et al. Functional interaction between BLM helicase and 53BP1 in a Chk1-mediated pathway during S-phase arrest. *J. Cell Biol.* 166, 801–813 (2004).
23. Ward, I. et al. The tandem BRCT domain of 53BP1 is not required for its repair function. *J. Biol. Chem.* 281, 38472–38477 (2006).
24. Iwabuchi, K. et al. Stimulation of p53-mediated transcriptional activation by the p53-binding proteins, 53BP1 and 53BP2. *J. Biol. Chem.* 273, 26061–26068 (1998).
25. Brummelkamp, T. R. et al. An shRNA barcode screen provides insight into cancer cell vulnerability to MDM2 inhibitors. *Nature Chem. Biol.* 2, 202–206 (2006).
26. Toledo, F. & Wahl, G. M. Regulating the p53 pathway: in vitro hypotheses, in vivo veritas. *Nature Rev. Cancer* 6, 909–923 (2006).
27. Morales, J. C. et al. 53BP1 and p53 synergize to suppress genomic instability and lymphomagenesis. *Proc. Natl Acad. Sci. USA* 103, 3310–3315 (2006).

Supplementary Information is linked to the online version of the paper at www.nature.com/nature.

Acknowledgements We thank N. Barlev for the Set9 expression vector; D. Reinberg for the p53K372me1 antibody; S. Benchimol for the BJ and BJ-DNp53 cell lines; T. Halazonetis for 53BP1 plasmid; R. Schule for p21 luciferase vector; and members of the T.J. and S.L.B. laboratories for discussions. This project is funded, in part, by a AACR-Pennsylvania Department of Health Fellows grant and Leukemia and Lymphoma Society Special Fellow grant (J.H.). M.T.B. is supported by a Welch Foundation grant. Research in the laboratory of T.J. is supported by the IMP through Boehringer Ingelheim and by grants from the European Union and the Austrian GEN-AU initiative, which is financed by the Austrian Ministry of Education, Science and Culture. Research support to S.L.B. was provided by a grant from the National Cancer Institute at NIH and the Commonwealth Universal Research Enhancement Program of the Pennsylvania Department of Health.

Author Contributions J.H., R.S., A.B.E., M.G.L., J.A.D., M.R. and S.O. performed the experimental work; R.S., M.T.B., T.J. and S.L.B. were responsible for project planning and data analysis.

Author Information Reprints and permissions information is available at www.nature.com/reprints. The authors declare no competing financial interests. Correspondence and requests for materials should be addressed to S.L.B. (berger@wistar.org).

Supplementary information:

Methods

Antibodies

Anti-Smyd2 (amino terminus) and anti-p53K370me1, me2 and me3 antibodies were generated and described previously[2]. Anti-p53K370me2 antibody was antigen-purified from crude serum. Other antibodies were as follows: anti-p53K372me1 (gift from D. Reinberg), anti-LSD1 (Bethyl), anti-p53 (DO1; Santa Cruz), anti-p53 (FL393; Santa Cruz), anti-haemagglutinin (Roche), anti-GST (Upstate), anti-53BP1 (monoclonal; Upstate), anti-53BP1 (polyclonal; Bethyl), anti-β-actin (Sigma), anti-Puma N terminus (Sigma), anti-Flag (Sigma) and mouse and rabbit IgG (Santa Cruz).

Co-immunoprecipitation assay

In the standard co-immunoprecipitation assay, cells were lysed in NET 0.1% buffer (50 mM Tris-HCl pH 7.4, 150 mM NaCl, 5 mM EDTA, 0.1% Nonidet P40, freshly added 1 mM phenylmethylsulphonyl fluoride, protease inhibitors) and sonicated with a Bioruptor (Diagenode) for 5 min, with 30-s 'on' and 1-min 'off' cycles. After removal of cell debris by centrifugation, 2 μg of antibody was added to 1 mg of clarified whole cell extract (WCE) and incubated overnight at 4 °C. The next day, 40 μl of Protein A-agarose beads (Upstate) was added and incubated for a further 2 h. Beads then were subjected to three washes with NET 0.1% buffer and boiled with 1× SDS loading buffer. A modified co-immunoprecipitation assay was used to detect interaction between p53 and 53BP1, in which nuclei were prepared in accordance with the protocol from a Nuclear Extraction Kit (catalogue no. AY2002; Panomics). Nuclear extract (NE) was prepared by lysing the nuclei with NET 0.1% supplemented with 2 mM sodium orthovanadate, 50 mM NaF and 50 mM β-glycerolphosphate as described above for WCE. Typically, 1 mg of NE (3–5 μg μl^{-1}) was used for each immunoprecipitation with 2 μg of specific antibody. For Flag immunoprecipitation, to assess the binding of Flag–53BP1 to K370me1 and K370me2 *in vivo*, *Smyd2* and *p53* were coexpressed with or without Flag–53BP1. The same amounts of input and Flag immunoprecipitation eluates were loaded in western blot analysis for the detection of K370me1 and K370me2.

GST pull-down assay

Bacterial lysate for GST–p53 or GST was pre-bound to 10 μl of GST beads (Pharmacia) at 25 °C. Beads were then preblocked with *in vitro* pull-down buffer (20 mM HEPES pH 7.9, 150 mM KCl, 1 mM EDTA, 0.1% Nonidet P40, 10% glycerol, 1 μg μl^{-1} BSA and freshly added 1 mM dithreitol and 1 mM phenylmethylsulphonyl fluoride) at 4 °C for 1 h. Recombinant Flag–LSD1 (2 μg) purified from baculovirus-infected *Spodoptera frugiperda* (Sf9) cells was added and incubated for a further 3 h. Beads were washed three times with pull-down buffer, then boiled with 20 μl of 1 × SDS loading buffer. Anti-Flag antibody was used to detect Flag–LSD1.

RNA interference

siRNAs were purchased from Dharmacon, and 100 nM siRNA was transfected into cells with DharmaFECT 1. siRNA sequences for luciferase control were as described previously[2]. LSD1 siRNA sequences were:

5'-UGAAUUAGCUGAAACACAAUU-3' (sense sequence) and

5'-pUUGUGUUUCAGCUAAUUCAUU-3' (antisense sequence).

Lentivirus-based shRNA vectors for luciferase control and LSD1 were generated with the BLOCK-it U6 RNAi Expression System (Invitrogen). shRNA vectors for 53BP1#1 and 53BP1#2 were purchased from Sigma. The sequences of the short hairpins were: luciferase control,

5'-CACCTAAGGCTATGAAGAGATACCGAAGTATCTCTTCATAGCCTTA-3' (top strand) and

5'-AAAATAAGGCTATGAAGAGATACTTCGGTATCTCTTCATAGCCTTA-3' (bottom strand);

LSD1,

5'-CACCGCACCTTATAACAGTGATACTCGAAAGTATCACTGTTATAAGGTGC-3'

(top strand) and

5'-AAAAGCACCTTATAACAGTGATACTTTCGAGTATCACTGTTATAAGGTGC-3' (bottom strand);

53BP1#1,

5'-CCGGGATACTTGGTCTTACTGGTTTCTCGAGAAACCAGTAAGACCAAGTATCTTTTT-3' (top strand) and
5'-AAAAAGATACTTGGTCTTACTGGTTTCTCGAGAAACCAGTAAGACCAAGTATCCCGG-3' (bottom strand);
53BP1#2,
5'-CCGGCCAGTGTGATTAGTATTGATTCTCGAGAATCAATACTAATCACACTGGTTTTT-3' (top strand) and
5'-AAAAACCAGTGTGATTAGTATTGATTCTCGAGAATCAATACTAATCACACTGGCCGG-3' (bottom strand).

We found that lentivirus-based knockdown of LSD1 by shRNA was not stable and the level of LSD1 recovered two weeks after transduction (data not shown). To overcome this technical difficulty, we repeated the cell transduction two weeks after the first transduction. By this 'double transduction' procedure, we were able to maintain a low level of LSD1 in U2OS, BJ and BJ-DNp53 cells for a month. All the experiments were performed within a month of the first transduction.

Chromatin immunoprecipitation assay

In brief, cells were fixed with 1% formaldehyde and lysed with lysis buffer (50 mM HEPES-KOH pH 7.5, 140 mM NaCl, 1 mM EDTA, 0.1% Triton X-100, 0.1% sodium deoxycholate, with protease inhibitors). The cell lysate was sonicated with a Bioruptor machine for a total of 20 min with 30-s 'on' and 1-min 'off' cycles, to shear the DNA to a final size of 200–500 base pairs. After preclearing with Protein A or G beads (Upstate), antibody was added and incubated overnight at 4 °C. The next day, Protein A-agarose or Protein G-agarose beads were added and incubated for a further 1–2 h. The complex was washed twice with lysis buffer, once with high salt buffer (50 mM HEPES-KOH pH 7.5, 500 mM NaCl, 1 mM EDTA, 0.1% Triton X-100, 0.1% sodium deoxycholate), twice with LiCl buffer (10 mM Tris-HCl pH 8.0, 0.25 M LiCl, 0.5% Nonidet P40, 0.5% sodium deoxycholate, 1 mM EDTA) and once with TE buffer, followed by elution in TE buffer containing 1% SDS. The crosslinks were reversed, the DNA was purified with a QIAquick 8 PCR purification kit (Qiagen) and subjected to analysis by quantitative real-time PCR.

In vitro demethylation assay

Recombinant Set9, GST–p53 and LSD1 were purified as described previously[1, 5, 11]. Recombinant Smyd2 was purified from insect (Sf9) cells infected with baculovirus. Methylation assays were performed as described previously[2]. GST–p53 (10 μg) and Set9 (5 μg) or Smyd2 (5 μg) were used in a 50-μl methylation reaction. After passage of the methylation reaction products through a Microspin column 30 (Bio-Rad), 1 μl of flowthrough was subjected to scintillation counting to measure the level of methylated GST–p53. Comparable scintillation counts within the flowthroughs were used in the demethylation assays with 3–5 μg of LSD1 in demethylation buffer (50 mM HEPES/NaOH pH 8.0, 25% glycerol)[4, 5, 28]. For peptide demethylation, 100 pmol of peptide and 5 μg of LSD1 purified from bacteria were used in a 50 μl total volume. After reaction, products were subjected to dot-blot analysis with specific antibodies.

Dual luciferase assay

Dual luciferase assays were performed in accordance with the manufacturer's protocol (Promega). In brief, H1299 cells were seeded in a 24-well plate. At 16 h after transfection, cells were lysed with passive lysis buffer and lysates were subjected to dual luciferase analysis with a Wallac 1420 multilabel counter (PerkinElmer). A *Renilla* luciferase vector (pRL-CMV) was used as an internal control.

Protein domain microarray

Chromo, PhD, Tudor, SANT, MBT and PWWP domains from about 100 chromatin-associated proteins were fused to GST and spotted on a slide[18]. Biotin-labelled p53K370me0 (unmethylated), K370me1, K370me2 and K370me3 peptides were individually hybridized to the protein domain array; binding was then detected with Cy5-Streptavidin[18].

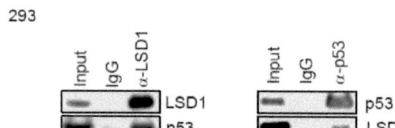

Figure S1. LSD1 interacts with p53 in 293 cells. Co-immunopreciptiation assay was used to examine interaction between endogenous LSD1 and p53.

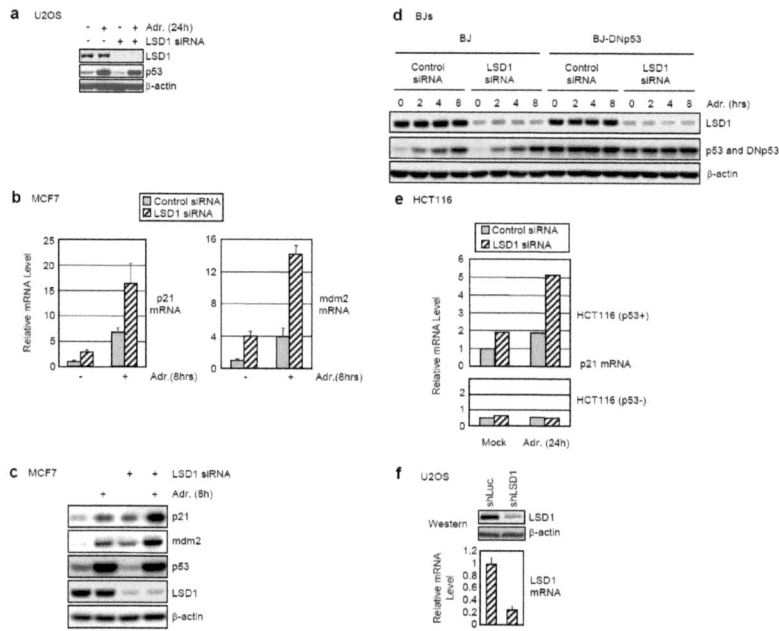

Figure S2. LSD1 represses p53's activity. a) Proteins levels of LSD1, p53 and β-actin in U2OS cells were analyzed by Western blotting. **b)** Relative mRNA levels of p21 and mdm2 in MCF7 cells were examined by quantitative realtime PCR (q-PCR) **c)** Western blot analysis was used to assay protein levels of p21, mdm2, p53, LSD1 and β-actin. **d)** Western blot analysis was used to assay protein levels of LSD1, p53 and β-actin in BJ and BJ-DNp53 cells. **e)** Q-PCR was used to determine relative mRNA levels of p21 in HCT116 (p53+/+) and HCT116 (p53-/-) cells. **f)** Western blot analysis was used to assay protein (upper panels) and q-PCR was used to assay relative mRNA (lower panels) of LSD1 in U2OS cells. Relative mRNA levels were calculated by normalizing absolute RNA level to GAPDH RNA level. Results were mean ± s.d.

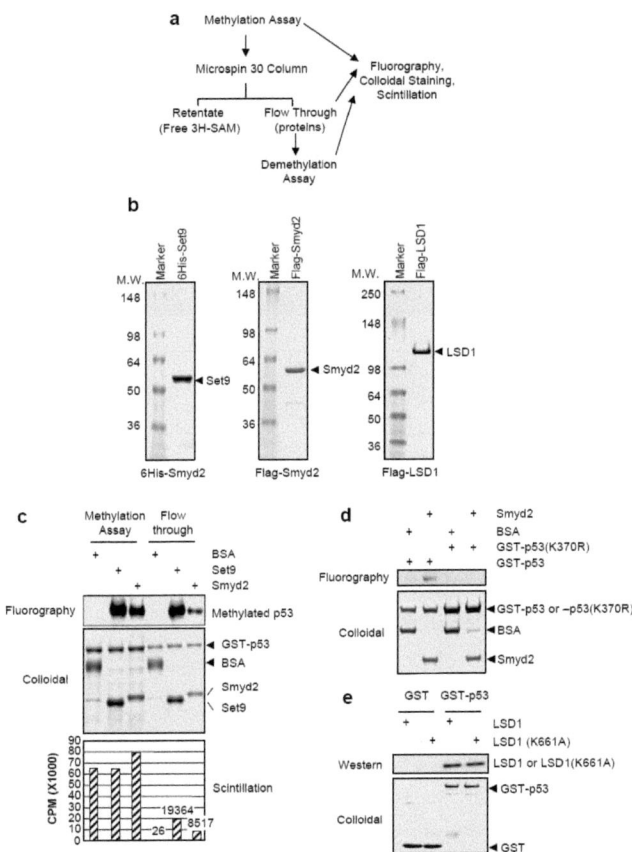

Figure S3. LSD1 demethylates p53 at K370. a) Schematic of *in vitro* demethylation assay. **b)** Colloidal staining of recombinant proteins purified from bacteria and insect cells. 6-Histidine tagged Set9 was purified from bacteria. Flag tagged LSD1 and Smyd2 were purified from insect (sf9) cells infected with baculovirus. **c)** Methylation of GST-p53 by Set9 and Smyd2 were examined by fluorography before or after column (upper panel). Colloidal staining shows equal loading of proteins (middle panels). Scintillation counts correspond to methylation of GST-p53 by Set9 and Smyd2 before or after column (bottom panels). **d)** Fluorography of wild type p53 or p53K370R to examine Smyd2 methylation. **e)** GST pull-down assay to assess the binding of recombinant LSD1 and LSD1(K661A) to p53.

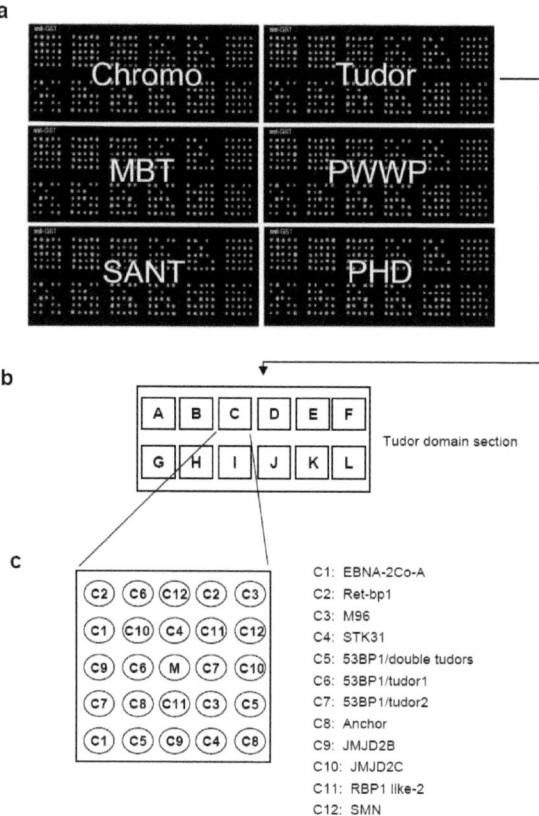

Figure S4. Schematic of GST protein domain array. a) Grouping of the entire GST protein domain array according to the functional domains. **b)** Schematic of tudor domain section of GST protein domain array. **c)** Schematic of C3 subsection where the tandem tudor domains of 53BP1 is fabricated.

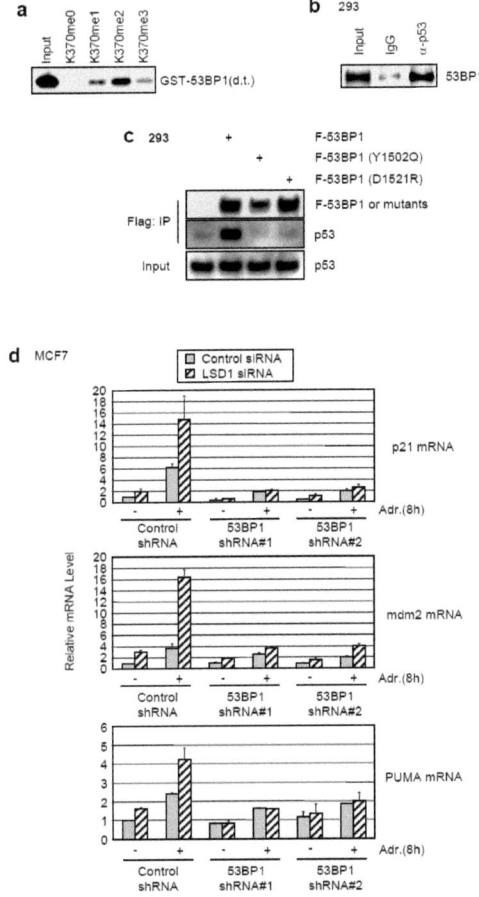

Figure S5. LSD1 represses p53's activity through 53BP1. a) Peptide pull-down assessment of tandem tudor domains of 53BP1 binding to p53K370me0, me1, me2 and me3 peptides. **b)** Interaction between endogenous p53 and 53BP1 in 293 cell was determined by coimmunoprecipitation, as described in Fig. S1a. **c)** Flag-IP assay was used to examine the role of the tandem tudor domains of 53BP1 in interaction between p53 and 53BP1. **d)** Q-PCR analysis of relative mRNA levels of p21, mdm2 and PUMA in MCF7 cells.

4.4. Discussion

Our results demonstrate two new possibilities for the regulation of p53 showing parallels with general chromatin biology and histone lysine methylation. One is that methylation at one lysine residue can have two different outputs depending on the number of methyl-groups added. Lysine 370 mono-methylation by Smyd2 is repressive while di-methylation is activating for p53 function. For histone lysine-methylation this concept has already been described. H4 lysine 20 mono-methylation for example is found at sites of active transcription while tri-methylation is associated with silenced regions of the genome. The second is the possibility of cross-talk between methylation at distinct residues. Set9 mediated activating methylation at K372 inhibits repressive methylation at K370 by Smyd2. A similar example is histone H3 lysine 9 tri-methylation as this mark when associated with pericentric heterochromatin catalyzed by Suv39h1 and h2 is needed for the induction of H4K20 tri-methylation by Suv4-20h1 and h2. p53 being such an important molecule for the protection of genomic integrity has to be tightly regulated and has to respond to different stimuli in a well defined output. This is in perfect agreement with the plethora of posttranslational modifications (PTMs) found on this molecule. Our in vivo results clearly indicate that methylation is amongst those regulating modifications and can be used to fine-tune p53 activity. Whether this possibility is ubiquitously used has to be answered by experiments on knockout animals for the enzymes or such carrying point mutations at the lysine residues modified in the p53 gene. In a study investigating the phenotype of mice in which the C-terminal 7 lysine residues have been replaced by arginines it was found that they seemed phenotypically normal. The only minor differences were the resistance of their fibroblasts (MEFs) to spontaneous immortalization and a better activation of p53 function after irradiation of thymocytes. This study can not answer the question for the specific function of the modifying enzyme or the impact of a defined lysine residue but it highlights that the PTMs are not generally needed for p53 activity but are important for the integration of stimuli or stresses leading to a well defined output.

To better understand this integration it would be interesting to investigate how the enzymes are controlled and whether they are capable of activating or inhibiting other factors. This could allow the establishment of communication networks within the cells and could thereby help to identify novel approaches in tumor therapy not targeting such an important molecule as p53 directly but interfering with its regulation. An example is that broad histone deacetylase inhibitors such as trichostatin A (TSA) prolong the half-life of p53 (Ito et al. 2001; Cheng et al. 2003) or in case of depsipeptide increase the transcription of its target genes (Zhao et al. 2006). Our study also shows that although extensive research has been carried out on this important regulator its function is still not fully elucidated and more work is needed to understand its potential for therapeutic treatment.

5. Materials and Methods

For buffer composition, peptide sequences, primer sequences and plasmid maps see appendix.

Bioinformatic analysis of murine SET domain containing enzymes:

Using the Ensembl database (www.ensembl.org) and the Simple Modular Architecture Research Tool (SMART) we extracted the sequence of all identified SET domains in the murine genome. We analyzed them using the ClustalX software package for their similarity and clustered them accordingly. This data was then extracted and used for the construction of a phylogenetic tree using the online phylogenetic tree plotter of the Wageningen University (http://www.bioinformatics.nl/tools/plottree.html). Candidates for further analysis were chosen according to this data.

Production of recombinant enzymes:

cDNAs of candidate genes from the SET- domain tree where subcloned into GATEWAY (Invitrogen) compatible pGEX6.1 vector (GE Healthcare Lifesciences) and transformed into BL21-CodonPlus(DE3)-RP or RIL strain. A 50mL overnight culture (37°C) in lysogeny broth (LB) medium was diluted next day in 500mL of fresh medium and cultured to an optical density (OD600) of 1.0. Protein expression was induced with 1mM Isopropyl β-D-1-thiogalactopyranoside (IPTG). Cells were cultured at 25°C for an additional 3-6 hours and then harvested at 4400rpm for 20 minutes. The cell pellet was resuspended in 30 mL phosphate buffered saline (PBS) and transferred to a 50mL Falcon. After a second centrifugation step the pellet was frozen in liquid nitrogen and resuspended in 30mL active lysis buffer and rotated for 90 minutes at 4°C.

For reduction of protease activity a Complete, EDTA-free; protease inhibitor cocktail tablet (Roche Applied Science) was added at this time. The suspension was sonicated five times for 15 seconds with a 30 second break on ice between each cycle using a microtip on a Branson Sonifier (S-450A/ 50% duty cycle/

output 6) followed by centrifugation in a SS34 rotor (10 minutes/ 12.000g/ 4°C) to remove insoluble debris.
The cleared lysate was used for protein purification.

Purification of recombinant enzymes:

The glutathione S-transferase (GST) tagged proteins were purified from the cleared bacterial lysate in the following way. 500µL of Glutathione Sepharose 4B (GE Healthcare Lifesciences) beads were prepared as in the manufacturer's manual and washed with cold PBS containing 1mM Dithiothreitol (DTT). The beads were transferred to the lysate and incubated rotating at 4°C for 2 hours. After incubation the beads were collected by centrifugation (5 minutes/ 500g/ 4°C) and washed with 30mL of cold PBS containing 1mM DTT by rotation at 4°C for 5 minutes. Washing step was repeated once. The beads were then collected in poly-prep disposable columns (Bio-Rad). Residual buffer was removed on a vacuum manifold (QIAvac 24 Plus; Qiagen). The recombinant protein was cleaved of the GST-tag using PreScission protease (GE Healtcare Lifesciences). 20µL of PreScission protease were mixed with 980µL of PreScission protease buffer and transferred to the beads. The mixture was incubated overnight at 4°C.
The following day 2mL of PrecScission protease buffer were added and the recombinant protein was collected in a 15mL Falcon tube by centrifugation (2 minutes/ 500g/ 4°C). The buffer was exchanged for PBS containing 10% glycerol using Amicon Ultra-4 centrifugal filter units (Millipore) using an appropriate molecular weight cut-off. The protein was analyzed on a sodium dodecyl sulfate polyacrylamide gel electrophoresis (SDS-PAGE / 12%) stained with Coomassie Brilliant Blue R-250 (peqlab) and snap frozen in liquid nitrogen before transferred to -80°C for long-term storage.

Production of recombinant histones:

The cDNAs for the *Xenopus laevis* core histones were subcloned into the pET3a (New England Biolabs (NEB)) expression vector and transformed into the BL21-Gold (DE3)pLysS (Stratagene) *E.coli* strain. A 5mL overnight culture (37°C) in LB medium was diluted next day in 500mL fresh medium and cultured to an OD600 of 0.5. After the induction with 0.5mM IPTG protein was expressed at 37°C for 3 hours. The cells were collected by centrifugation (20minutes/ 4400rpm/4°C), resuspended in 30mL of WASH buffer and snap frozen in liquid nitrogen. The suspension was thawed at room temperature and sonicated 3 times for 30 seconds with a 30 seconds break on ice between each cycle using a microtip on a Branson Sonifier (S-450A/ 50% duty cycle/ output 6) followed by centrifugation in a SS34 rotor (20 minutes/ 17,000rpm/ 4°C) to collect inclusion bodies. The pellet was washed 3 times in 30mL WASH buffer with 1% Triton X-100 (Sigma-Aldrich) and 3 times in WASH buffer without detergent. For improved resuspension the samples were sonicated between the washing steps for a few pulses using the same settings as given above. The inclusion bodies were collected each time by centrifugation (20 minutes/ 4,400rpm/ 4°C). After the last step the pellet was resuspended in 3-4 mL of UNFOLDING buffer and rotated at RT for 1 hour followed by several pulses of sonication (settings see above). The suspension was rotated for another 15 minutes at RT. To remove insoluble material the sample was centrifuged in a SS34 rotor (5 minutes/ 17,000rpm/ 4°C). The soluble fraction contained the purified histones. Quality and concentration of the preparation was checked on a 15% SDS-PAGE using bovine serum albumin (BSA) at defined concentrations as standard. Point mutations were introduced using the QuikChange site-directed mutagenesis kit (Stratagene).

Histone methyltransferase assay

For radioactive assays tritium labeled S-adenosyl methionine was used (S-Adenosyl-l-methyl-^{3}H]methionine (^{3}H-SAM); GE Healthcare Lifesciences). 1µg of recombinant enzyme was mixed with 1µg of recombinant histone and 1.5µL of ^{3}H-SAM (1µCi/µL) in a 50mM Tris-HCl pH 8.5 buffer containing 5mM DTT. The total volume of the reaction was 25µL. For peptide based assays 800µM of the peptide were used. The mixture was incubated for 2 hours at 30°C and run on a 15% (histones) or 12%(peptides) SDS-PAGE gel. For peptides the gels were incubated in ENLIGHTNING (Perkin Elmer) rapid autoradiography enhancer for 30 minutes and dried using a vacuum gel-dryer before overnight exposure to Amersham Hyperfilm MP (GE Healthcare Lifesciences). For histones the gels were blotted to Immobilon-P membrane (Millipore) using a TE70 semi-dry transfer unit (GE Healthcare Lifesciences) and sprayed 3 times with EN3HANCE spray (Perkin Elmer) according to manual before overnight exposure to Amersham Hyperfilm MP. The time of exposure was adapted to signal intensity.

In a non-radioactive assay on peptides using matrix-assisted laser desorption/ionization (MALDI) mass spectrometry for detection of the methylated form of the peptide concentrations ranging from 80µM to 0.8µM for the peptide were used. To determine the methylated positions the modified peptide was fragmented and MS/MS spectra were recorded allowing identifying the target lysine.

All peptides were dissolved in water at a stock concentration of 10mM.

Mass spectrometric analysis

The samples of the HMTase assay were diluted 1:50 with 0.1% trifluoroacetic acid (TFA) (Pierce) before spotting 1.5µL onto the MALDI sample plate. Each sample was spotted twice. 1µL of the matrix (2.2 mg/ml HCCA in 70% ACN, 0.1% TFA) was added and spots were dried at room temperature. Spectra were measured on a 4800 MALDI TOF/TOF (Applied Biosystems).

Expression analysis of Setd6 and Setd3 by northern blots

For Setd6 a 1kb fragment of the 1.4kb coding region was used for the production of the probe and for Setd3 the full coding region. After amplification by PCR probes labeled with [α−32P]dATP were generated using the Prime-It II Random Primer Labeling Kit (Stratagene). Those were subsequently purified from the reaction by illustra Sephadex G-50 (GE Healthcare Life Sciences) size exclusion spin columns. The incorporation of radioactivity was measured by Cerenkov counts. The northern blots BLOT3 (Sigma Aldrich), MTNI and II (Clontech) were pre-hybridized in 5mL of ExpressHyb (Clontech) solution at 68°C for 30 minutes. The probes were denatured at 95°C for 5 minutes, chilled quickly on ice and added to 5mL of fresh ExpressHyb solution. The membrane was incubated in the mixture for 1 hour at 68°C rotating. Per membrane 2×10^6 cpm/mL were used. After the incubation the blots were washed in 2xSSC/0.05%SDS several times and then continuously for 40 minutes on a shaker. In a second washing step 0.1xSSC/0.1%SDS at 50°C for 40 minutes were the used conditions. Excess solution was shaken off and the membrane was wrapped in plastic foil before exposure to a phosphor screen overnight at room temperature. The next day the screens were analyzed using a STORM 860 phosphorimager (GE Healthcare Lifesciences). The blots were incubated in 0.5%SDS (100°C) for 10 minutes shaking and allowed to cool down in the solution for stripping. They were stored at -20°C.

Production of polyclonal rabbit antibodies

Antibodies for Setd6 and Setd3 were raised against the full length recombinant protein after the removal of the GST-tag. For each antibody the pre-immune sera of 6 rabbits were tested and of these 3 were chosen for the production. Per rabbit a total of 700µg recombinant protein was used. The solution had a concentration of 0.5µg/µL. For the peptide antibodies against the methyl-marks 10µmol of branched peptide were used for the injection per rabbit after the coupling to keyhole limpet hemocyanin (KLH). Those antibodies were affinity purified.

Antibodies were made with the help of Gramsch laboratories (Schwabhausen; Germany) for injection, animal handling and extraction of the sera.

Immunofluorescence staining

Cells were plated on Lab-Tek II 8-well chamber slides (NUNC) usually one day in advance. For ES cells the chamber slides were coated using a sterile 0.2% gelatine solution in PBS. The next day the chambers were removed according to the manual and the cells were washed once in PBS using a coplin jar and then fixed in a 2% paraformaldehyde/ PBS solution. Subsequently they were washed twice for 5 minutes each time in PBS before the permeabilization with a 0.1% sodium citrate solution containing 0.1% of Triton X-100 (Sigma) for 5 minutes. Cells were washed again with PBS twice and then another two times in PBS containing 0.25% bovine serum albumine (BSA; Sigma) and 0.1% Tween-20 (Sigma) (5 minutes each wash). Before the incubation with the primary antibody the chamber slides were blocked with PBS/0.25%BSA/0.1%Tween-20 and 10% normal goat serum (JacksonImmuno) or rabbit serum for antibodies raised in goat for 30 minutes. Now they were incubated overnight at 4°C with the primary antibody in the blocking solution. For Setd6 and Setd3 the crude serum was diluted 1:1000. The methyl specific antibodies were affinity purified and used at a 1:1000 dilution. All other antibodies were used according to the recommendation of their vendor. After the overnight incubation the slides were washed 3 times 10 minutes at room temperature using PBS/0.25%BSA/0.1%Tween-20 before incubating with the secondary antibody in the blocking solution for 1 hour at room temperature. This was followed by 3 additional washing steps. 200µL of mounting media containing DAPI (4',6-Diamidin-2'-phenylindol- dihydrochlorid) was applied to each slide (Vector labs). The coverslips were fixed with nail polish subsequently.

Cell culture

NIH 3T3, w9 iMEFs (female wildtype immortalized mouse embryonic fibroblasts) and 293FT cells were cultured in DMEM-complete media. Male feeder-independent mouse embryonic stem cells (wt26) were grown in DMEM-ES and on tissue culture dishes that were coated with gelatine. All cell lines were split every 2 days in a 1:10 ratio.

Production of lentiviral particles

On day one 2-2.5x 10^6 293FT cells were plated per 10cm dish. The next day the cells were transfected using the lipofectamine and plus reagent (Invitrogen) in the following way. In a 1.5mL Eppendorf tube 3.33µg of the plasmid coding for the transgene or the shRNA were mixed with 2.5µg of psPAX2 and 1µg of pMD2.G in 750µL of DMEM-L-Glu. 20µL of plus reagent were slowly added and mixed well by pipetting. In a second 1.5mL tube 750µL DMEM-L-Glu were mixed well with 30µL of lipofectamine. Both solutions remained at room temperature for 15 minutes. After this period they were combined and incubated for another 15minutes. Meanwhile the medium of the cells was replaced with 5mL DMEM-L-Glu. After the incubation the DNA mix was added dropwise to cells distributing it evenly over the plate. The cells were left for 6 hours in the incubator (37°C/5%CO_2) before the mixture was aspirated off and exchanged for 6mL of DMEM-complete. From this time on all steps were performed according to biosafety level 2 regulations. 48 hours later the virus was harvested in the following way. The media containing the viral particles was transferred to a fresh 15mL conical Falcon tube and centrifuged for 5 minutes at 1500rpm to remove debris and loose cells. The supernatant was aliquoted into 1.5mL Eppendorf tubes (1mL each) and stored at -80°C until used for further experiments. psPAX2 contains all elements needed for packaging (gag/pol) with the exception of the envelope gene which is encoded on the pMD2G plasmid.

Western blotting using polyclonal rabbit antibodies

For the analysis of the reduction of Setd6 and the associated methyl-marks after knockdown whole cell extracts of confluent 6 well dishes were prepared. For this purpose cells were harvested by trypsinization and washed once in PBS. The cell pellet was resuspended in 100μL of 2x SDS loading dye and placed at 95°C for 5 minutes. To reduce the viscosity the samples were sonicated using a micro tip sonicator (Branson) for a few pulses (output 5/ 50% duty cycle) before loading on a 15% SDS-PAGE gel. The proteins were then blotted onto a Immobilon-P membrane using a Hoefer TE77 semi-dry transfer unit (GE Healthcare Lifesciences). After the transfer they were rinsed in 100% methanol and air-dried. The membranes were incubated in 5%Blotto (Santa Cruz Biotechnology)/1xTBS-T with the primary antibody at room temperature for 2 hours rotating followed by 3 consecutive washes in 1xTBS-T (10 minutes each). The Setd6 antibody (1452) was used a 1:1000 dilution, the H2AK5me antibody at 1:500, the H3K14me antibody at 1:2000 and all other methyl-specific antibodies at 1:1000. The membranes were incubated in 5%Blotto/1xTBS-T containing the secondary anti-rabbit HRP coupled antibody (1:2000; JacksonImmuno) for 1 hour at room temperature before washing as described before. For detection the ECL reagent was used (GE Healthcare Lifesciences).

Transcriptional analysis by cDNA microarrays

For the analysis we extracted total RNA of wildtype NIH 3T3, Setd6 shRNA 3 clone 9 (shRNA 3/9) and Setd6 shRNA 13 clone 11 (shRNA13/11) in duplicates using the trizol (Invitrogen) reagent. After a quality control of the RNA using the Agilent Bioanalyzer 2100 those RNA preparations were amplified using the MessageAmp II aRNA amplification kit (Ambion) according to the supplier's manual. The next step was the labelling with the Alexa dyes 555 or 647 after cDNA generation and incorporation of amino-allyl dUTP. The so generated samples were hybridized to the arrays using a Tecan HS4800 hybridization station. The slides were analyzed on an Axon 4000B using the GenePix Pro 6.0 software package (Molecular Devices) and a custom pipeline for quality control.

The results were extracted into an Access (Microsoft) database file and analyzed as indicated in the results.

Immunoprecipitation of Setd6 and identification of interaction partners

Cellular extracts from mouse NIH 3T3 or embryonic stem cells were prepared in the following way. The cells were harvested by scraping from the tissue culture dishes and collected by centrifugation for 5 minutes at 1500rpm in a 50mL Falcon tube. The pellet was washed with 30mL ice cold PBS and collected again by centrifugation. For extraction of the cytosolic proteins the pellet was resuspended in 2mL of buffer A per gram. NP40 was added to a concentration of 0.1% before passing the suspension 4 times through a 18G1 needle. The solution was incubated on ice for 10 minutes and spun at 4400rpm for 10 minutes at 4°C. The supernatant contains the major fraction of the cytosolic proteins. The pellet was resuspended in 2mL of buffer B per initial gram of cell pellet. Sodium chloride was added to a final concentration of 0.4M before passing the suspension 4 times through a 18G1 needle and incubating it for 30 minutes on ice. This was followed by centrifugation at 4400rpm for 20 minutes at 4°C. The supernatant contains the soluble fraction of the nuclear proteins. For the immunoprecipitation (IP) we used Dynabeads Protein A (Invitrogen) covalently linked to the rabbit polyclonal anti-Setd6 antibody (1452) using dimethyl pimelimidate dihydrochloride (DMP; Pierce). Per 100μL of beads we used 20μL of the crude serum. The preparation was done according to the manufacturer's manual. For each IP (per antibody) we used 4 grams of cell pellet and 100μL of Dynabeads for each fraction (cytosolic and nuclear). After cross-linking the beads were added to the extracts together with a Complete Mini protease inhibitor tablet (Roche Applied Science) followed by an overnight incubation at 4°C rotating. The next day the mixture was centrifuged at 1000rpm for 2 minutes and the beads were collected on a magnetic stand. Depending on the fraction the beads were washed in 8 mL of the initial lysis buffer (buffer A or B) for 15 minutes at 4°C rotating. They were collected on the magnetic stand and transferred to a 1.5mL Eppendorf tube and washed two more times in 1mL of lysis buffer rotating at 4°C.

Bound proteins were then eluted by 2 minute incubation with 40µL of 0.1M citrate pH 2.3. The pH of the eluate was increased by the addition of 150µL of 1M Tris/HCl pH8. 50µL of the sample was loaded onto a NuPAGE Novex 4-12% Bis-Tris gel (Invitrogen) and run using MOPS buffer. The gels were silver stained according to the Blum protocol. Unique bands were excised and given to mass spectrometry for identification. As control we used the regions of the gel in the beads-only or Setd3 sample at the same molecular weight.

Subcloning of shRNAs into pENTR H1/TO

1µg of pENTR H1/TO with insert was restricted overnight with 10 units BsmB1 (NEB) at 37°C and gel purified (20µL reaction). The linear plasmid was then dephosphorylated to prevent self-ligation by incubation with 5 units of Anarctic phosphatase (NEB) for 1 hour at 37°C. The forward and reverse strand primers containing the shRNAs were annealed by heating to 95°C and slowly cooling down to room temperature in a 40µL reaction containing each primer at a concentration of 25µM and 1x oligo annealing buffer. In the next step the primers were phosphorylated using T4 polynucleotide kinase (Roche Applied Science). 1µL of the annealed primers were used and the kinase was heat inactivated after a 30 minutes reaction at 37°C. Hereafter the primers were diluted 1:100 in distilled H_2O and 2µL were ligated with 1µL of the dephosphorylated plasmid in a 20µL reaction using T4 DNA ligase (Promega). The sample was incubated overnight at 16°C before transformation into DH5α and selection on agar plates containing 50µg/mL kanamycin. Colonies were picked and DNA was purified using the Qiagen QIAprep Spin Miniprep Kit. Positive insertion was checked by a diagnostic digest using EcoRV. In case of religation a 250bp band is visible. The DNA was then sequenced to check for a proper insertion. The shRNAs were then subcloned into pLenti4BlockItPuro for lentiviral packaging and expression in mammalian cells using the LR clonase II mix (Invitrogen).

6. Appendix

6.1. Buffer composition:

Phosphate buffered saline (PBS) per litre:
- 8g   Sodium chloride
- 0.2g   Potassium chloride
- 1.44g Na_2HPO_4
- 0.24g KH_2PO_4
- Adjust pH to 7.4
- fill up with distilled H_2O

Prescission protease buffer:
- 50mM   Tris-HCl
- 150mM   Sodium chloride
- 1mM   EDTA (ethylenediaminetetraacetic acid)
- 1mM   DTT (Dithiothreitol)
- Adjust pH to 7.0 at 25°C

SDS Running buffer (10x) per 2 litres:
- 60.4g   Tris-Base
- 288g   Glycine
- 200mL   10% SDS (sodium dodecyl sulphate) in H_2O
- fill up with distilled H_2O

Active lysis buffer:
- 0,5% NP40
- 1% Triton X-100
- 150mM NaCl
- 1mM PMSF
- 2mg/mL lysozyme (from hen egg white)
- 25U/mL Benzonase (or 10µL DNaseI + 10µL RNase T1 in 1 L lysis buffer)
- in MonoQ

4x SDS loading dye per 100mL:
- 8g SDS
- 46mL Glycerol (87%)
- 1.54g DTT
- 24mL 1M Tris pH6.8
- 4mg Bromophenol blue
- fill up with distilled H_2O

Tris bufferd saline with Tween 20 (10x) per 2 litres:
- 1L 1M Tris pH8
- 400mL 5M Sodium chloride
- 20mL Tween 20
- fill up with distilled H_2O

SDS-gel staining buffer:
- 50% Methanol
- 10% Acetic acid
- 40% H_2O
- 0.25% Coomassie brilliant blue R-250

SDS-gel destaining buffer:
- 50%   Methanol
- 10%   Acetic acid
- 40% H_2O

Oligo annealing buffer (10x):
- 100mM   Tris-HCl pH8.0
- 10mM   EDTA pH8.0
- 1M   NaCl

Buffer A:
- 10 mM Hepes pH 7.9
- 5 mM $MgCl_2$
- 0.25 M Sucrose
- add protease inhibitors before use
- filter sterilize and store at 4 °C

Buffer B:
- 10 mM Hepes pH 7.9
- 1 mM $MgCl_2$
- 0.1 mM EDTA
- 25% glycerol
- add protease inhibitors before use
- filter sterilize and store at 4 °C

MOPS based running buffer for Novex gradient gels (20x):
- 1M   Tris Base
- 1M   MOPS (4-Morpholinepropanesulfonic acid)
- 20.5mM   EDTA
- 69.3mM   SDS

Tris acetate EDTA buffer (10x) for DNA gels:
- 0.4M   Tris Base
- 0.2M   Acetic acid
- 0.01M   EDTA

WASH buffer for recombinant histones:
- 50mM   Tris/HCl pH7.5
- 100mM   NaCl
- 1mM   EDTA
- 1 Complete protease inhibitor tablet (Roche Applied Science) per 50mL

UNFOLDING buffer for recombinant histones:
- 7M   Guanidine hydrochloride
- 20mM   Tris/HCl pH7.5
- 10mM   DTT

Washing solution for immunofluorescence:
- 0.25%   Bovine serum albumine
- 0.1%   Tween 20
- in 1x PBS

Blocking solution for immunofluorescence:
- Washing solution with 10% normal goat serum (JacksonImmuno)

6.2. Media composition and concentration of antibiotics:

Lysogeny broth (LB) medium per litre:
- 10g   Tryptone
- 5g   Yeast extract
- 10g   NaCl
- dissolve in distilled water and autoclave

DMEM-complete:
- 440mL   High glucose Dulbecco's Modified Eagle Medium (DMEM)
- 50mL   Fetal calf serum (FCS)
- 5mL   Penicillin Streptomycin solution (Invitrogen) (100x)
- 5mL   L-glutamine (200mM)

DMEM-ES:
- 403mL   High glucose Dulbecco's Modified Eagle Medium (DMEM)
- 75mL   FCS ES cell culture tested
- 5mL   Penicillin Streptomycin solution (Invitrogen) (100x)
- 5mL   L-glutamine (200mM)
- 5mL   MEM Non essential amino acids (Invitrogen) (100x)
- 5mL   MEM Sodium pyruvate (Invitrogen) (100mM)
- 1mL   β-Mercaptoethanol (50mM)
- 1.25mL   Filter-sterilizied supernatant of Leukemia inhibitory factor (LIF) producing cells

Antibiotics for bacterial cell culture:
Ampicillin   100μg/mL
Chloramphenicol   170μg/mL
Kanamycin   50μg/mL
Zeocin   50μg/mL

Antibiotics for mammalian cell culture:

G418 300μg/mL

Puromycin 2μg/mL for ES and 2.5μg/mL for NIH 3T3 and w9

6.3. Peptide sequences:

Substrate	Region	Sequence
H1e	AA 18-37	TPVKKKARKAAGGAKRKTSG
H2A	AA 1-20	SGRGKQGGKARAKAKTRSS
H3	AA 1-30	ARTKQTARKSTGGKAPRKQLATKAARKSA
H4	AA 1-28	SGRGKGGKGLGKGGAKRHRKVLRDNIQ
Sgsm2	AA 112-124	QGSTGGKAPALSP
Pin4	AA 5-17	GKSGSGKGGKGGA
Wbscr17	AA 89-102	GYGGRGKGGLPAT
Col17a1	AA 433-443	GGGRGKGGGAG
Zfp369	AA 619-631	KTSGRGKGRKINA
Gpr98	AA 5092-5104	ETRGLGKGGVNWR
Suclg1	AA 289-301	AIIAGGKGGAKEK
Smarca2	AA 998-1010	EKDKKGKGGAKTL
H2AK5me	For antibody	SGRGK(me)QGGKARA
H3K14me	For antibody	KSTGGK(me)APRKQL
H4K5me	For antibody	SGRGK(me)GGKGLG
H4K12me	For antibody	GKGLGK(me)GGAKRH

6.4. Primer sequences

Name	Descripiton	Forward
Setd6gw Forward	Subcloning of cDNA	GGGGACAAGTTTGTACAAAAAAGCAGGCTTAACC ATGGCGGCCCCCGCCAAGCGCGCGCGGGTAAG
Setd6gw Reverse	Subcloning of cDNA	GGGGACCACTTTGTACAAGAAAGCTGGGTC TTAATTTGTGAGTTCCAACACCCGATGTAAG
Setd3gw Forward	Subcloning of cDNA	GGGGACAAGTTTGTACAAAAAAGCAGGCTTAACC ATGGGTAAGAAGAGTCGAGTGAAAACTCAG
Setd3gw Reverse	Subcloning of cDNA	GGGGACCACTTTGTACAAGAAAGCTGGGTC TCAAAGTCGCTCCTTCACCTTGGCCATGC
Setd6RT Forward	For RT-PCR	CAGATGGTGACAGTCCGAGA
Setd6RT Reverse	For RT-PCR	CAGTCAGCACCTCCTCACAA
shRNA3 Forward	Insert for pEntrH1/TO	CACCGCAACTGATTCATATGTATGGCGAACCATACATATGAATCAGTTGC
shRNA3 Reverse	Insert for pEntrH1/TO	AAAAGCAACTGATTCATATGTATGGTTCGCCATACATATGAATCAGTTGC
shRNA9 Forward	Insert for pEntrH1/TO	CACCGGCTAACTGGCAACTGATTCACGAATGAATCAGTTGCCAGTTAGCC
shRNA9 Reverse	Insert for pEntrH1/TO	AAAAGGCTAACTGGCAACTGATTCATTCGTGAATCAGTTGCCAGTTAGCC
shRNA13 Forward	Insert for pEntrH1/TO	CACCGCAGACATACTAAACCACATACTCGAGTATGTGGTTTAGTATGTCTGC
shRNA13 Reverse	Insert for pEntrH1/TO	AAAAGCAGACATACTAAACCACATACTCGAGTATGTGGTTTAGTATGTCTGC
Setd6mut Forward	Y285A	GAAATCTTCAACACAGCCGGGCAAATGGCTAAC
Setd6mut Reverse	Y285A	GTTAGCCATTTGCCCGGCTGTGTTGAAGATTTC
Setd3mut Forward	Y313A	CAGATTTACATTTTTGCCGGCACTCGGTCAAATG
Setd3mut Reverse	Y313A	CATTTGACCGAGTGCCGGCAAAAATGTAAATCTG
Setd6 RT primer1F	Northern probe	CTGTTCCATCAGCGGCCTGCTGGAG
Setd6 RT primer3R	Northern probe	GGCATAGGCCTCCTTATTACTGAG
Setd3north Forward	Northern probe	ATGGGTAAGAAGAGTCGAGTG
Setd3north Reverse	Northern probe	AAGTCGCTCCTTCACCTTG

Sequences in red are either needed for subcloning into plasmids or introduce mutations into the original coding sequence. All sequences are written from 5' to 3' end

6.5. Plasmid maps:

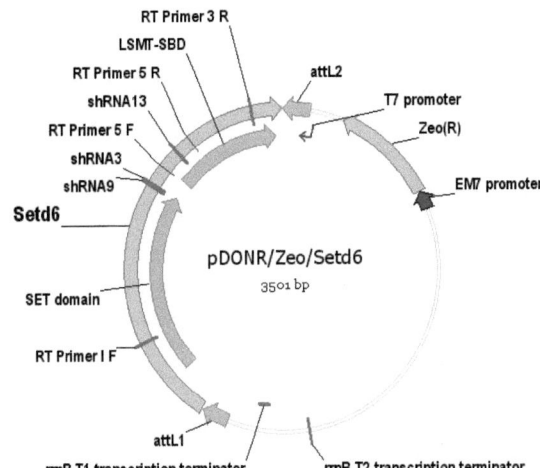

pDonr/Zeo/Setd6 contains the full length cDNA of Setd6.

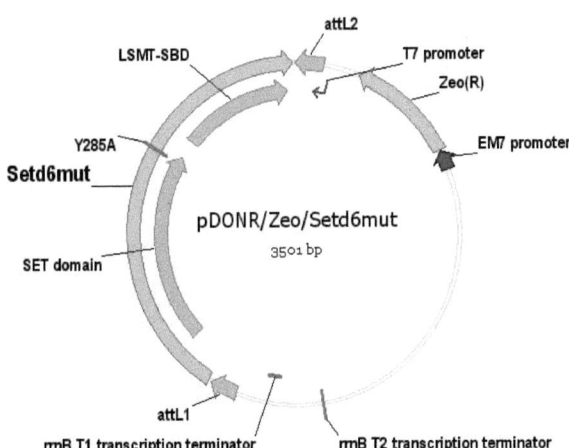

pDonr/Zeo/Setd6mut contains the full length cDNA of Setd6 with a tyrosine-alanine substitution at position 285.

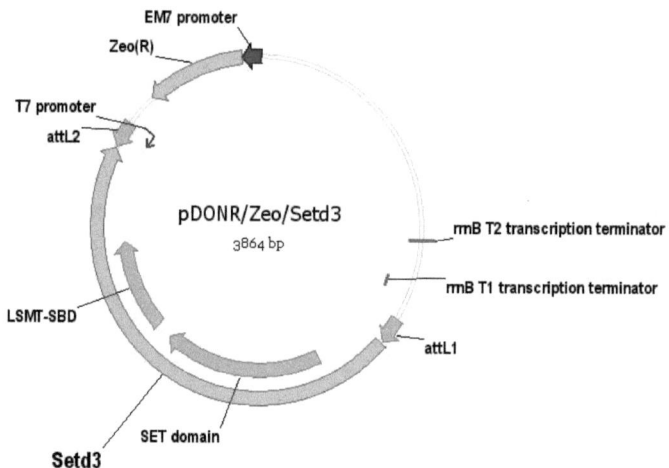

pDONR/Zeo/Setd3 contains the full length cDNA of Setd3

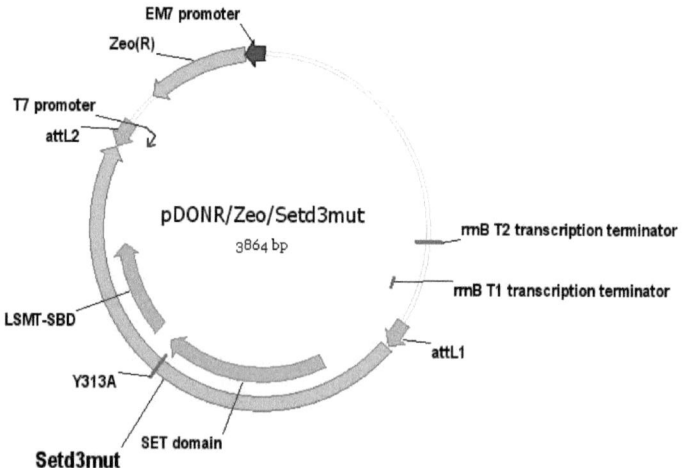

pDONR/Zeo/Setd3mut contains the full length cDNA of Setd3 with a tyrosine to alanine substitution at position 313

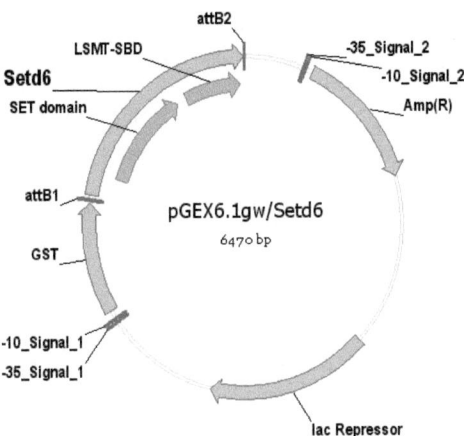

pGEX6.1gw/Setd6 was used for the expression of a GST-fusion of Setd6 in bacteria. The GST-tag can be cleaved of using Prescission protease

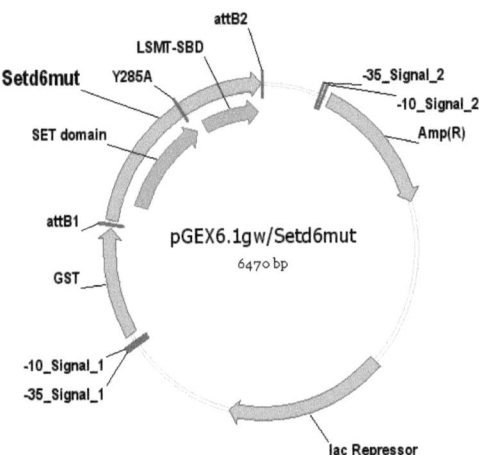

pGEX6.1gw/Setd6mut was used for the expression of a GST-fusion of the mutant Setd6 in bacteria. The GST-tag can be cleaved of using Prescission protease

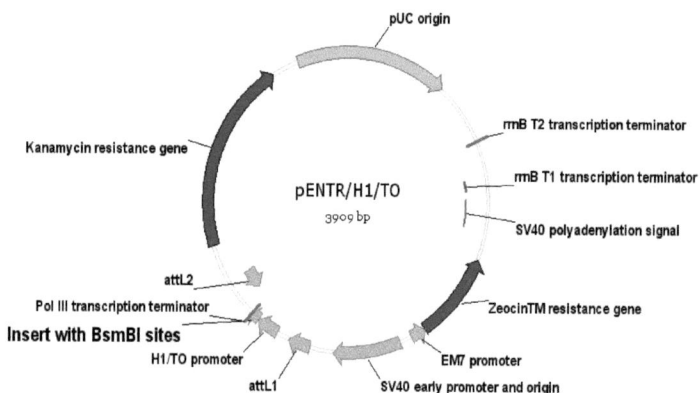

pENTR/H1/TO was used for the first subcloning step of the shRNAs. Once cloned into this plasmid they could be easily subcloned into lentiviral destination plasmids.

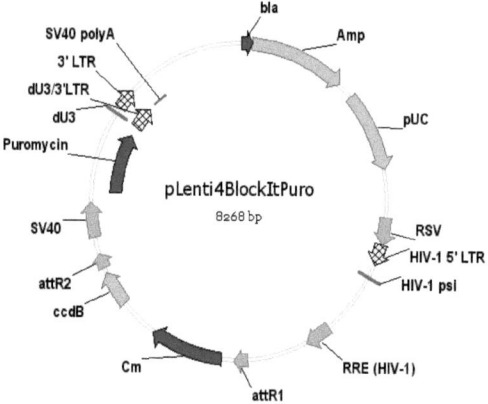

pLenti4BlockItPuro was used for the transduction of mammalian cells to stably express the subcloned shRNAs. Puromycin can be used as resistance marker.

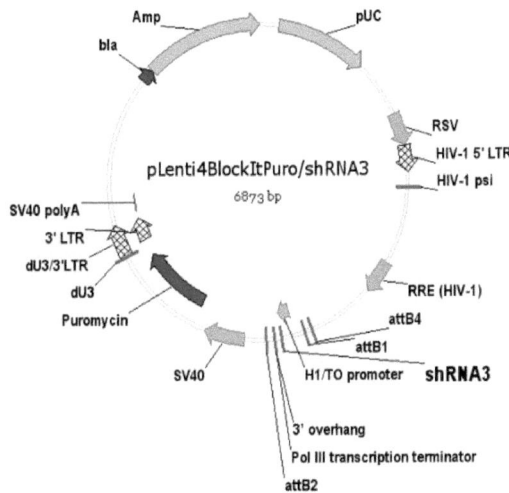

pLenti4BlockItPuro/shRNA3 is an example for the final plasmid used for expression of shRNAs in cells. The shRNA is transcribed from the H1 promoter using Pol III.

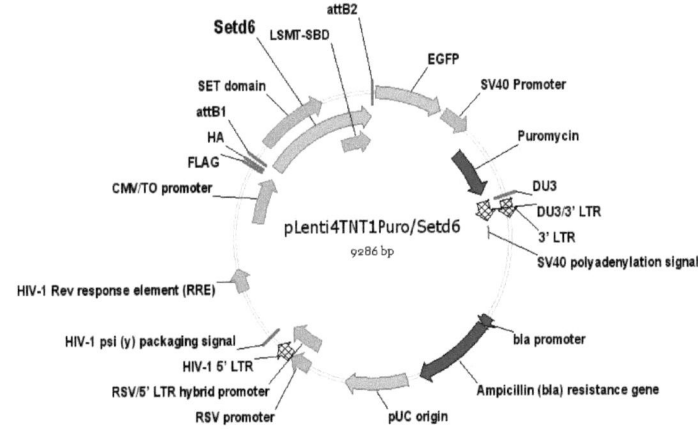

pLenti4TNT1Puro/Setd6 encodes a Flag-HA-Set6-EGFP fusion protein for overexpression studies in mammalian cells using lentiviral transduction.

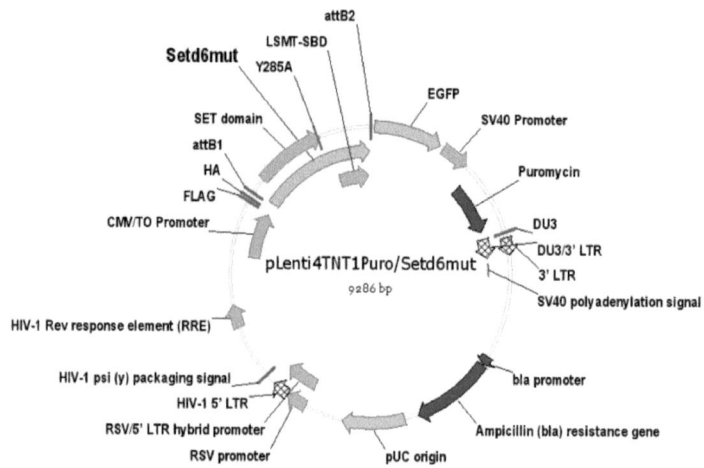

pLenti4TNT1Puro/Setd6mut encodes a Flag-HA-Set6mut-EGFP fusion protein for overexpression studies in mammalian cells using lentiviral transduction.

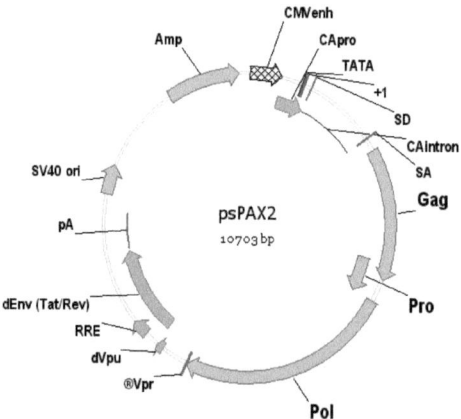

psPAX2 contains the genes needed for lentiviral packaging and was co-transfected with the destination vectors and pMD2.G in 293FT cells.

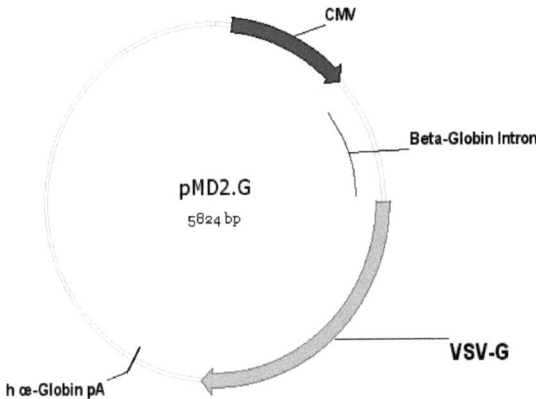

pMD2.G is used for the expression of the envelope gene needed for production of viral particles. It was co-transfected with psPAX2 and the appropriate destination vector into 293FT cells for packaging.

6.6. Dot-blots

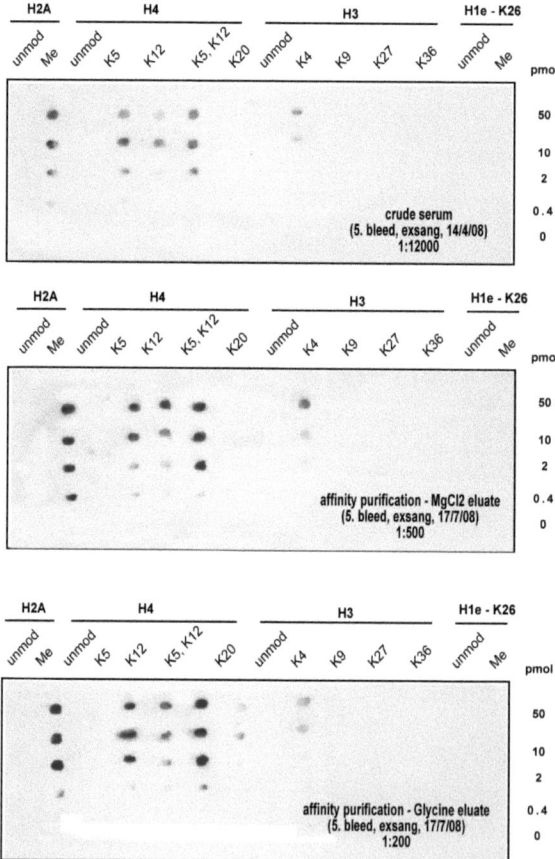

Figure 6-1 Shown are dot-blot for α-H2AK5me1 #1919 crude serum and the affinity purified antibody eluted with MgCl$_2$ or glycine

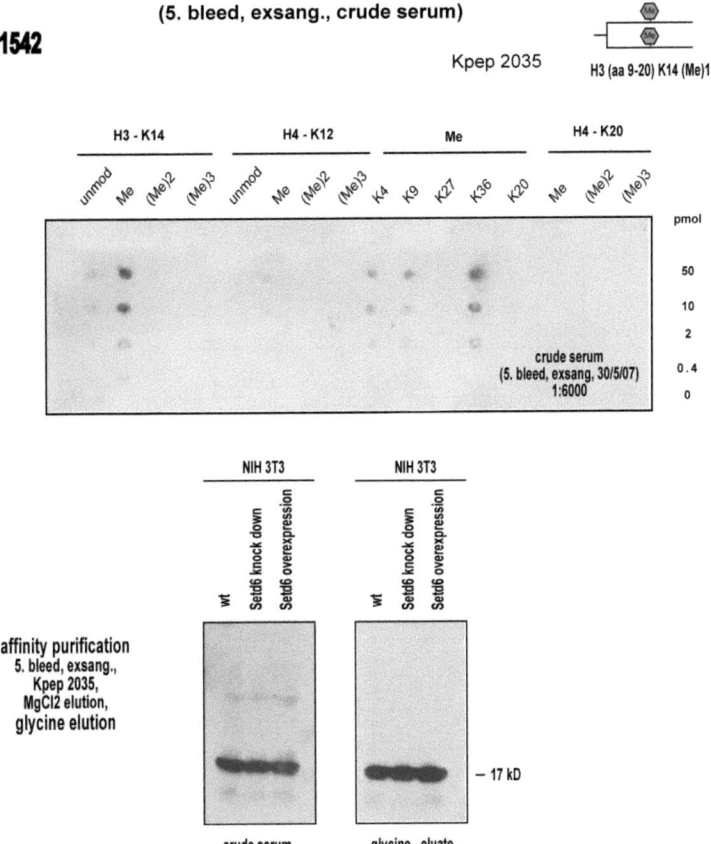

Figure 6-2 Shown are the dot-blot for α-H3K14me1 #1542 crude serum and the affinity purified antibody eluted with $MgCl_2$ or glycine used for western blots on whole cell extracts from murine NIH 3T3 cells

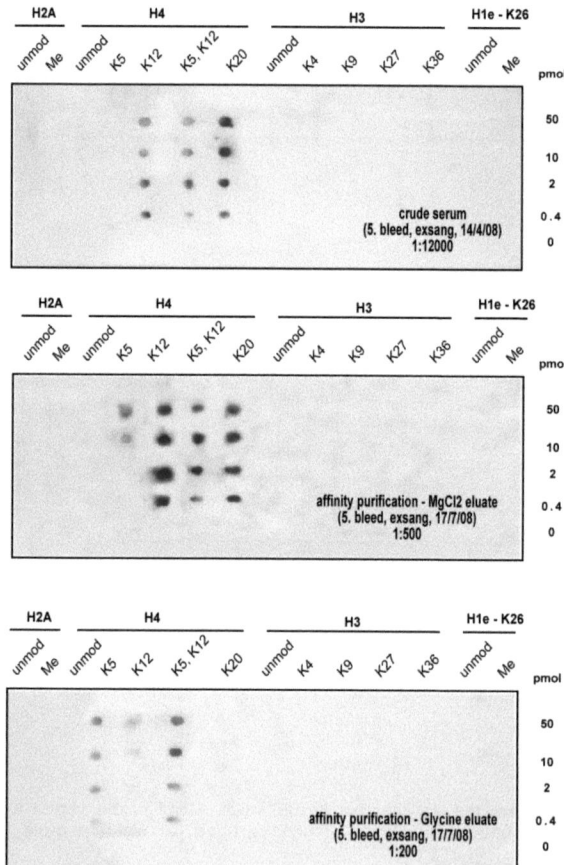

Figure 6-3 Shown are dot-blot for α-H4K5me1 #1910 crude serum and the affinity purified antibody eluted with $MgCl_2$ or glycine

α-2b H4 K12 me1 antibodies

#1543 (5. bleed/exsang., crude serum)   Kpep 2036

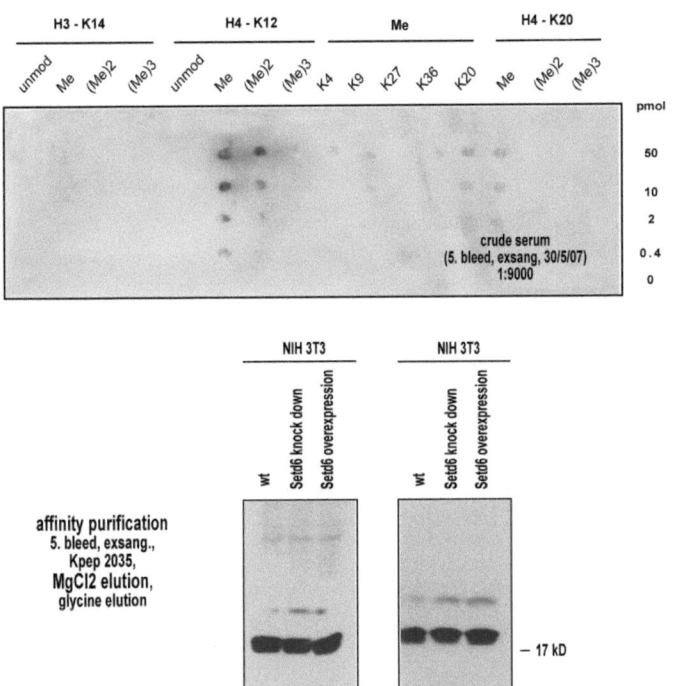

Figure 6-4 Shown are the dot-blot for α-H3K14me1 #1543 crude serum and the affinity purified antibody eluted with MgCl₂ or glycine used for western blots on whole cell extracts from murine NIH 3T3 cells

7. References

Aagaard, L., G. Laible, et al. (1999). "Functional mammalian homologues of the Drosophila PEV-modifier Su(var)3-9 encode centromere-associated proteins which complex with the heterochromatin component M31." Embo J **18**(7): 1923-38.

Allis, C. D., Jenuwein, T., Reinberg, D (2007). Epigenetics, Cold Spring Harbor Laboratory Press.

Aoki, M., H. Kiyonari, et al. (2008). "R-spondin2 expression in the apical ectodermal ridge is essential for outgrowth and patterning in mouse limb development." Dev Growth Differ **50**(2): 85-95.

Beisel, C., A. Buness, et al. (2007). "Comparing active and repressed expression states of genes controlled by the Polycomb/Trithorax group proteins." Proc Natl Acad Sci U S A **104**(42): 16615-20.

Benjamin, J. M. and W. J. Nelson (2008). "Bench to bedside and back again: molecular mechanisms of alpha-catenin function and roles in tumorigenesis." Semin Cancer Biol **18**(1): 53-64.

Benzel, I., Y. A. Barde, et al. (2001). "Strain-specific complementation between NRIF1 and NRIF2, two zinc finger proteins sharing structural and biochemical properties." Gene **281**(1-2): 19-30.

Berger, S. L. (1999). "Gene activation by histone and factor acetyltransferases." Curr Opin Cell Biol **11**(3): 336-41.

Berger, S. L. (2007). "The complex language of chromatin regulation during transcription." Nature **447**(7143): 407-12.

Bode, A. M. and Z. Dong (2004). "Post-translational modification of p53 in tumorigenesis." Nat Rev Cancer **4**(10): 793-805.

Bonisch, C., S. M. Nieratschker, et al. (2008). "Chromatin proteomics and epigenetic regulatory circuits." Expert Rev Proteomics **5**(1): 105-19.

Cang, Y., J. Zhang, et al. (2006). "Deletion of DDB1 in mouse brain and lens leads to p53-dependent elimination of proliferating cells." Cell **127**(5): 929-40.

Cang, Y., J. Zhang, et al. (2007). "DDB1 is essential for genomic stability in developing epidermis." Proc Natl Acad Sci U S A **104**(8): 2733-7.

Carninci, P., T. Kasukawa, et al. (2005). "The transcriptional landscape of the mammalian genome." Science **309**(5740): 1559-63.

Carrozza, M. J., B. Li, et al. (2005). "Histone H3 methylation by Set2 directs deacetylation of coding regions by Rpd3S to suppress spurious intragenic transcription." Cell **123**(4): 581-92.

Cheng, H. L., R. Mostoslavsky, et al. (2003). "Developmental defects and p53 hyperacetylation in Sir2 homolog (SIRT1)-deficient mice." Proc Natl Acad Sci U S A **100**(19): 10794-9.

Chin, H. G., P. O. Esteve, et al. (2007). "Automethylation of G9a and its implication in wider substrate specificity and HP1 binding." Nucleic Acids Res **35**(21): 7313-23.

Chuikov, S., J. K. Kurash, et al. (2004). "Regulation of p53 activity through lysine methylation." Nature **432**(7015): 353-60.

Clegg, H. V., K. Itahana, et al. (2008). "Unlocking the Mdm2-p53 loop: ubiquitin is the key." Cell Cycle **7**(3): 287-92.

Coisy-Quivy, M., O. Disson, et al. (2006). "Role for Brm in cell growth control." Cancer Res **66**(10): 5069-76.

Cuddapah, S., R. Jothi, et al. (2009). "Global analysis of the insulator binding protein CTCF in chromatin barrier regions reveals demarcation of active and repressive domains." Genome Res **19**(1): 24-32.

Deng, C. X. (2009). "SIRT1, is it a tumor promoter or tumor suppressor?" Int J Biol Sci **5**(2): 147-52.

Erhardt, S., I. H. Su, et al. (2003). "Consequences of the depletion of zygotic and embryonic enhancer of zeste 2 during preimplantation mouse development." Development **130**(18): 4235-48.

Fan, Y., T. Nikitina, et al. (2003). "H1 linker histones are essential for mouse development and affect nucleosome spacing in vivo." Mol Cell Biol **23**(13): 4559-72.

Fazzio, T. G., J. T. Huff, et al. (2008). "An RNAi screen of chromatin proteins identifies Tip60-p400 as a regulator of embryonic stem cell identity." Cell **134**(1): 162-74.

Feng, Y., N. Lee, et al. (2003). "TIP49 regulates beta-catenin-mediated neoplastic transformation and T-cell factor target gene induction via effects on chromatin remodeling." Cancer Res **63**(24): 8726-34.

Glozak, M. A., N. Sengupta, et al. (2005). "Acetylation and deacetylation of non-histone proteins." Gene **363**: 15-23.

Gottifredi, V. and C. Prives (2001). "P53 and PML: new partners in tumor suppression." Trends Cell Biol **11**(5): 184-7.

Greig, K. T., S. Carotta, et al. (2008). "Critical roles for c-Myb in hematopoietic progenitor cells." Semin Immunol **20**(4): 247-56.

Hayashi, K. and Y. Matsui (2006). "Meisetz, a novel histone tri-methyltransferase, regulates meiosis-specific epigenesis." Cell Cycle **5**(6): 615-20.

Hayashizaki, Y. and P. Carninci (2006). "Genome Network and FANTOM3: assessing the complexity of the transcriptome." PLoS Genet **2**(4): e63.

He, L., J. X. Yu, et al. (1998). "RIZ1, but not the alternative RIZ2 product of the same gene, is underexpressed in breast cancer, and forced RIZ1 expression causes G2-M cell cycle arrest and/or apoptosis." Cancer Res **58**(19): 4238-44.

Huang, J. and S. L. Berger (2008). "The emerging field of dynamic lysine methylation of non-histone proteins." Curr Opin Genet Dev **18**(2): 152-8.

Huang, S., G. Shao, et al. (1998). "The PR domain of the Rb-binding zinc finger protein RIZ1 is a protein binding interface and is related to the SET domain functioning in chromatin-mediated gene expression." J Biol Chem **273**(26): 15933-9.

Huen, M. S., S. M. Sy, et al. (2008). "Direct interaction between SET8 and proliferating cell nuclear antigen couples H4-K20 methylation with DNA replication." J Biol Chem **283**(17): 11073-7.

Huh, W. K., J. V. Falvo, et al. (2003). "Global analysis of protein localization in budding yeast." Nature **425**(6959): 686-91.

Iijima, K., M. Ohara, et al. (2008). "Dancing on damaged chromatin: functions of ATM and the RAD50/MRE11/NBS1 complex in cellular responses to DNA damage." J Radiat Res (Tokyo) **49**(5): 451-64.

Imai, S. and H. Kitano (1998). "Heterochromatin islands and their dynamic reorganization: a hypothesis for three distinctive features of cellular aging." Exp Gerontol 33(6): 555-70.
Ito, A., C. H. Lai, et al. (2001). "p300/CBP-mediated p53 acetylation is commonly induced by p53-activating agents and inhibited by MDM2." Embo J 20(6): 1331-40.
Itoh, T., K. Miyake, et al. (2008). "Differentiation-specific expression of chromatin remodeling factor BRM." Biochem Biophys Res Commun 366(3): 827-33.
Ivanov, G. S., T. Ivanova, et al. (2007). "Methylation-acetylation interplay activates p53 in response to DNA damage." Mol Cell Biol 27(19): 6756-69.
Jenuwein, T. (2001). "Re-SET-ting heterochromatin by histone methyltransferases." Trends Cell Biol 11(6): 266-73.
Jenuwein, T. and C. D. Allis (2001). "Translating the histone code." Science 293(5532): 1074-80.
Jenuwein, T., G. Laible, et al. (1998). "SET domain proteins modulate chromatin domains in eu- and heterochromatin." Cell Mol Life Sci 54(1): 80-93.
Kanei-Ishii, C., J. Ninomiya-Tsuji, et al. (2004). "Wnt-1 signal induces phosphorylation and degradation of c-Myb protein via TAK1, HIPK2, and NLK." Genes Dev 18(7): 816-29.
Karachentsev, D., K. Sarma, et al. (2005). "PR-Set7-dependent methylation of histone H4 Lys 20 functions in repression of gene expression and is essential for mitosis." Genes Dev 19(4): 431-5.
Kim, K. C., L. Geng, et al. (2003). "Inactivation of a histone methyltransferase by mutations in human cancers." Cancer Res 63(22): 7619-23.
Kim, M. Y., E. J. Ann, et al. (2007). "Tip60 histone acetyltransferase acts as a negative regulator of Notch1 signaling by means of acetylation." Mol Cell Biol 27(18): 6506-19.
Kimura, A. and M. Horikoshi (1998). "Tip60 acetylates six lysines of a specific class in core histones in vitro." Genes Cells 3(12): 789-800.
Kirmizis, A., S. M. Bartley, et al. (2004). "Silencing of human polycomb target genes is associated with methylation of histone H3 Lys 27." Genes Dev 18(13): 1592-605.
Klose, R. J., K. Yamane, et al. (2006). "The transcriptional repressor JHDM3A demethylates trimethyl histone H3 lysine 9 and lysine 36." Nature 442(7100): 312-6.
Kobayashi, J., K. Iwabuchi, et al. (2008). "Current topics in DNA double-strand break repair." J Radiat Res (Tokyo) 49(2): 93-103.
Lachner, M. and T. Jenuwein (2002). "The many faces of histone lysine methylation." Curr Opin Cell Biol 14(3): 286-98.
Latham, J. A. and S. Y. Dent (2007). "Cross-regulation of histone modifications." Nat Struct Mol Biol 14(11): 1017-1024.
Lawrence, M. C., A. Jivan, et al. (2008). "The roles of MAPKs in disease." Cell Res 18(4): 436-42.
Li, B., L. Howe, et al. (2003). "The Set2 histone methyltransferase functions through the phosphorylated carboxyl-terminal domain of RNA polymerase II." J Biol Chem 278(11): 8897-903.

Li, J., Q. E. Wang, et al. (2006). "DNA damage binding protein component DDB1 participates in nucleotide excision repair through DDB2 DNA-binding and cullin 4A ubiquitin ligase activity." Cancer Res **66**(17): 8590-7.

Loyola, A., T. Bonaldi, et al. (2006). "PTMs on H3 variants before chromatin assembly potentiate their final epigenetic state." Mol Cell **24**(2): 309-16.

Maeda, N., T. Kasukawa, et al. (2006). "Transcript annotation in FANTOM3: mouse gene catalog based on physical cDNAs." PLoS Genet **2**(4): e62.

Maresca, T. J., B. S. Freedman, et al. (2005). "Histone H1 is essential for mitotic chromosome architecture and segregation in Xenopus laevis egg extracts." J Cell Biol **169**(6): 859-69.

Merla, G., C. Ucla, et al. (2002). "Identification of additional transcripts in the Williams-Beuren syndrome critical region." Hum Genet **110**(5): 429-38.

Nakamura, N., S. Toba, et al. (2005). "Cloning and expression of a brain-specific putative UDP-GalNAc: polypeptide N-acetylgalactosaminyltransferase gene." Biol Pharm Bull **28**(3): 429-33.

Nam, J. S., E. Park, et al. (2007). "Mouse R-spondin2 is required for apical ectodermal ridge maintenance in the hindlimb." Dev Biol **311**(1): 124-35.

Nowak, S. J. and V. G. Corces (2004). "Phosphorylation of histone H3: a balancing act between chromosome condensation and transcriptional activation." Trends Genet **20**(4): 214-20.

O'Neill, L. P., A. M. Keohane, et al. (1999). "A developmental switch in H4 acetylation upstream of Xist plays a role in X chromosome inactivation." Embo J **18**(10): 2897-907.

Papp, B. and J. Muller (2006). "Histone trimethylation and the maintenance of transcriptional ON and OFF states by trxG and PcG proteins." Genes Dev **20**(15): 2041-54.

Perez-Burgos, L., A. H. Peters, et al. (2004). "Generation and characterization of methyl-lysine histone antibodies." Methods Enzymol **376**: 234-54.

Pfeifer, G. P. and G. P. Holmquist (1997). "Mutagenesis in the P53 gene." Biochim Biophys Acta **1333**(1): M1-8.

Pokutta, S., F. Drees, et al. (2008). "Biochemical and structural analysis of alpha-catenin in cell-cell contacts." Biochem Soc Trans **36**(Pt 2): 141-7.

Porras-Yakushi, T. R., J. P. Whitelegge, et al. (2006). "A novel SET domain methyltransferase in yeast: Rkm2-dependent trimethylation of ribosomal protein L12ab at lysine 10." J Biol Chem **281**(47): 35835-45.

Porras-Yakushi, T. R., J. P. Whitelegge, et al. (2005). "A novel SET domain methyltransferase modifies ribosomal protein Rpl23ab in yeast." J Biol Chem **280**(41): 34590-8.

Rathert, P., A. Dhayalan, et al. (2008). "Protein lysine methyltransferase G9a acts on non-histone targets." Nat Chem Biol **4**(6): 344-6.

Rea, S., F. Eisenhaber, et al. (2000). "Regulation of chromatin structure by site-specific histone H3 methyltransferases." Nature **406**(6796): 593-9.

Schiltz, R. L., C. A. Mizzen, et al. (1999). "Overlapping but distinct patterns of histone acetylation by the human coactivators p300 and PCAF within nucleosomal substrates." J Biol Chem **274**(3): 1189-92.

Schneider, R., A. J. Bannister, et al. (2004). "Histone H3 lysine 4 methylation patterns in higher eukaryotic genes." Nat Cell Biol 6(1): 73-7.

Schotta, G., A. Ebert, et al. (2002). "Central role of Drosophila SU(VAR)3-9 in histone H3-K9 methylation and heterochromatic gene silencing." Embo J 21(5): 1121-31.

Schotta, G., M. Lachner, et al. (2004). "A silencing pathway to induce H3-K9 and H4-K20 trimethylation at constitutive heterochromatin." Genes Dev 18(11): 1251-62.

Schotta, G., R. Sengupta, et al. (2008). "A chromatin-wide transition to H4K20 monomethylation impairs genome integrity and programmed DNA rearrangements in the mouse." Genes Dev 22(15): 2048-61.

Shi, X., T. Hong, et al. (2006). "ING2 PHD domain links histone H3 lysine 4 methylation to active gene repression." Nature 442(7098): 96-9.

Shi, X., I. Kachirskaia, et al. (2007). "Modulation of p53 function by SET8-mediated methylation at lysine 382." Mol Cell 27(4): 636-46.

Shilatifard, A. (2008). "Molecular implementation and physiological roles for histone H3 lysine 4 (H3K4) methylation." Curr Opin Cell Biol 20(3): 341-8.

Shukla, A., N. Stanojevic, et al. (2006). "Functional analysis of H2B-Lys-123 ubiquitination in regulation of H3-Lys-4 methylation and recruitment of RNA polymerase II at the coding sequences of several active genes in vivo." J Biol Chem 281(28): 19045-54.

Sims, R. J., 3rd and D. Reinberg (2008). "Is there a code embedded in proteins that is based on post-translational modifications?" Nat Rev Mol Cell Biol 9(10): 815-20.

Skradski, S. L., A. M. Clark, et al. (2001). "A novel gene causing a mendelian audiogenic mouse epilepsy." Neuron 31(4): 537-44.

Squatrito, M., C. Gorrini, et al. (2006). "Tip60 in DNA damage response and growth control: many tricks in one HAT." Trends Cell Biol 16(9): 433-42.

Steger, D. J., M. I. Lefterova, et al. (2008). "DOT1L/KMT4 recruitment and H3K79 methylation are ubiquitously coupled with gene transcription in mammalian cells." Mol Cell Biol 28(8): 2825-39.

Sterner, D. E. and S. L. Berger (2000). "Acetylation of histones and transcription-related factors." Microbiol Mol Biol Rev 64(2): 435-59.

Suganuma, T. and J. L. Workman (2008). "Crosstalk among Histone Modifications." Cell 135(4): 604-7.

Takeuchi, T., Y. Watanabe, et al. (2006). "Roles of jumonji and jumonji family genes in chromatin regulation and development." Dev Dyn 235(9): 2449-59.

Toledo, F. and G. M. Wahl (2006). "Regulating the p53 pathway: in vitro hypotheses, in vivo veritas." Nat Rev Cancer 6(12): 909-23.

Trievel, R. C., E. M. Flynn, et al. (2003). "Mechanism of multiple lysine methylation by the SET domain enzyme Rubisco LSMT." Nat Struct Biol 10(7): 545-52.

Trojer, P., G. Li, et al. (2007). "L3MBTL1, a histone-methylation-dependent chromatin lock." Cell 129(5): 915-28.

Tschiersch, B., A. Hofmann, et al. (1994). "The protein encoded by the Drosophila position-effect variegation suppressor gene Su(var)3-9 combines domains of antagonistic regulators of homeotic gene complexes." Embo J 13(16): 3822-31.

Turner, B. M. (2000). "Histone acetylation and an epigenetic code." Bioessays 22(9): 836-45.

Vaziri, H., S. K. Dessain, et al. (2001). "hSIR2(SIRT1) functions as an NAD-dependent p53 deacetylase." Cell **107**(2): 149-59.

Vincent-Salomon, A., C. Ganem-Elbaz, et al. (2007). "X inactive-specific transcript RNA coating and genetic instability of the X chromosome in BRCA1 breast tumors." Cancer Res **67**(11): 5134-40.

Walter, W., D. Clynes, et al. (2008). "14-3-3 interaction with histone H3 involves a dual modification pattern of phosphoacetylation." Mol Cell Biol **28**(8): 2840-9.

Wang, C. W., J. Kim, et al. (2001). "Apg2 is a novel protein required for the cytoplasm to vacuole targeting, autophagy, and pexophagy pathways." J Biol Chem **276**(32): 30442-51.

Watrin, E. and J. M. Peters (2006). "Cohesin and DNA damage repair." Exp Cell Res **312**(14): 2687-93.

Webb, K. J., A. Laganowsky, et al. (2008). "Identification of Two SET Domain Proteins Required for Methylation of Lysine Residues in Yeast Ribosomal Protein Rpl42ab." J Biol Chem **283**(51): 35561-8.

Wei, Q., C. Yokota, et al. (2007). "R-spondin1 is a high affinity ligand for LRP6 and induces LRP6 phosphorylation and beta-catenin signaling." J Biol Chem **282**(21): 15903-11.

Whetstine, J. R., A. Nottke, et al. (2006). "Reversal of histone lysine trimethylation by the JMJD2 family of histone demethylases." Cell **125**(3): 467-81.

Xie, M., G. Shao, et al. (1997). "Transcriptional repression mediated by the PR domain zinc finger gene RIZ." J Biol Chem **272**(42): 26360-6.

Yamamichi, N., K. Inada, et al. (2007). "Frequent loss of Brm expression in gastric cancer correlates with histologic features and differentiation state." Cancer Res **67**(22): 10727-35.

Yan, N. and Y. Shi (2003). "Histone H1.2 as a trigger for apoptosis." Nat Struct Biol **10**(12): 983-5.

Zhao, Y., S. Lu, et al. (2006). "Acetylation of p53 at lysine 373/382 by the histone deacetylase inhibitor depsipeptide induces expression of p21(Waf1/Cip1)." Mol Cell Biol **26**(7): 2782-90.

8. Acknowledgments

This thesis is dedicated to my family and especially my parents Karl and Margit Richter, who always supported my wishes and ideas. Without this constant support I would not have been able to reach this goal. They gave me the strength to continue in all the difficult situations.

The IMP for me was truly an inspiring place for my career. I met many people that in my scientific but also private life became good friends. I want to especially thank Inti De La Rosa-Velázquez, Barna Fodor and Georg Winter for a relaxed and motivating environment, but also the other past and current members of the Jenuwein lab. Thomas Jenuwein gave me the freedom to develop my own thoughts and helped me to stay focussed on the important aspects of my research. I also want to thank the members of my PhD committee Ortrun Mittelsten Scheid and Meinrad Busslinger for critical and inspiring discussions. I am grateful for the support that I received throughout these years from all the service departments at the IMP. Andreas Schmidt and Karl Mechtler helped me with all the mass spectrometry which was critical for the success of this work and were always open for discussions helping me to streamline my project. I hope that the friendships that started at this institute will continue and that many more exciting experiences together will follow.

I also want to thank all my friends outside the institute for supporting me during this time and for their understanding for my scientist's way of life.

Die VDM Verlagsservicegesellschaft sucht für wissenschaftliche Verlage abgeschlossene und herausragende

Dissertationen, Habilitationen, Diplomarbeiten, Master Theses, Magisterarbeiten usw.

für die kostenlose Publikation als Fachbuch.

Sie verfügen über eine Arbeit, die hohen inhaltlichen und formalen Ansprüchen genügt, und haben Interesse an einer honorarvergüteten Publikation?

Dann senden Sie bitte erste Informationen über sich und Ihre Arbeit per Email an *info@vdm-vsg.de*.

Sie erhalten kurzfristig unser Feedback!

VDM Verlagsservicegesellschaft mbH
Dudweiler Landstr. 99
D - 66123 Saarbrücken

Telefon +49 681 3720 174
Fax   +49 681 3720 1749

www.vdm-vsg.de

Die VDM Verlagsservicegesellschaft mbH vertritt

Printed by Books on Demand GmbH, Norderstedt / Germany